国家现代职业教育改革创新示范区建设成果
国家职业教育质量发展研究中心研发成果
工程实践创新项目（EPIP）教学模式规划教材

液压与气动技术

主编◎宫晓凯　刘恩宇

中国铁道出版社有限公司
CHINA RAILWAY PUBLISHING HOUSE CO., LTD.

内 容 简 介

本书主要包括液压元件、液压基本回路、气动元件、气动基本回路及其应用等内容，共7个项目23个任务。为提高学生兴趣和学习效率，在每个任务中都明确了任务目标。本书立足综合培养学生的动手能力，突出工学结合，注重分析和解决实际问题，具有一定的科学性和先进性。

本书实用性强，适合高等职业院校机械自动化类专业师生选用，也可作为液压与气动技术的培训教学参考书。

图书在版编目(CIP)数据

液压与气动技术：汉文、英文/宫晓凯，刘恩宇主编．—北京：中国铁道出版社有限公司，2023.7
国家现代职业教育改革创新示范区建设成果　国家职业教育质量发展研究中心研发成果
ISBN 978-7-113-30243-6

Ⅰ．①液… Ⅱ．①宫… ②刘… Ⅲ．①液压传动-双语教学-高等职业教育-教材-汉、英 ②气压传动-双语教学-高等职业教育-教材-汉、英 Ⅳ．①TH137 ②TH138

中国国家版本馆 CIP 数据核字(2023)第 084825 号

书　　名：	液压与气动技术（中英双语版）
作　　者：	宫晓凯　刘恩宇
策　　划：	何红艳　　　　　编辑部电话：(010)63560043
责任编辑：	何红艳　徐盼欣
封面设计：	刘　颖
责任校对：	刘　畅
责任印制：	樊启鹏

出版发行：中国铁道出版社有限公司(100054，北京市西城区右安门西街8号)
网　　址：http://www.tdpress.com/51eds/

印　　刷：河北京平诚乾印刷有限公司
版　　次：2023年7月第1版　2023年7月第1次印刷
开　　本：787 mm×1 092 mm　1/16　印张：15.25　字数：340千
书　　号：ISBN 978-7-113-30243-6
定　　价：49.80元

版权所有　侵权必究

凡购买铁道版图书，如有印制质量问题，请与本社教材图书营销部联系调换。电话：(010)63550836
打击盗版举报电话：(010)63549461

前言

党的二十大报告在回顾新时代十年来我国取得的历史性成就、发生的历史性变革时指出:"我们实行更加积极主动的开放战略,构建面向全球的高标准自由贸易区网络,加快推进自由贸易试验区、海南自由贸易港建设,共建'一带一路'成为深受欢迎的国际公共产品和国际合作平台。"为推动共建"一带一路"高质量发展,扩大与"一带一路"沿线国家的职业教育合作,贯彻落实天津市启动实施的将优质职业教育成果输出国门与世界分享计划的要求,职业教育作为与制造业联系紧密的一种教育形式,正在发挥着举足轻重的作用。

天津机电职业技术学院在马达加斯加建立了"鲁班工坊",为了配合马达加斯加"鲁班工坊"的理论和实训教学,开展交流与合作,实现教育资源共享,提高中国职业教育的国际影响力,创新职业院校国际合作模式,输出我国职业教育优质资源,编写了本书。

本书以鲁班工坊的液压设备为载体,基于项目教学,可以作为高等职业院校装备制造大类专业的职业能力培训教材。本书内容丰富,资料翔实,着重培养学生的动手能力,工学结合,注重分析和解决实际问题,具有科学性和先进性,突出下列特点:

(1)考虑高等职业教育人才岗位特点,采用贴近工程实际的教学案例,注重技术应用能力培养,突出实用技术训练。

(2)着重分析液压、气压设备各类元件的工作原理、结构及其应用。

(3)液压与气动职能符号严格遵循国际标准。

(4)对于复杂的结构原理图采用三维演示,以便于学生接受。

(5)为提高学生学习兴趣和效率,在每个任务中明确了具体目标。

本书共设7个项目,分别是液压动力元件的认知、液压执行元件、液压控制阀、基本液压回路的安装与调试、气压传动基础、气动元件的认知和气动回路。教师可根据学生实际情况安排学时。学时安排应以满足学生学习知识和提高动手能力为目的。

本书由天津机电职业技术学院宫晓凯、刘恩宇任主编。编写分工为:宫晓凯编写项目1~项目4,刘恩宇编写项目5~项目7。本书在编写过程中得到了浙江天煌科技实业有限公司的大力支持和帮助,在此向其表示衷心的感谢。

由于水平有限,书中疏漏和不妥之处在所难免,恩请广大读者批评指正。

编 者

2023年2月

目录

项目1 液压动力元件的认知 001
  任务1 液压动力元件概述 001
  任务2 齿轮泵的认知 005
  任务3 叶片泵的认知 008
  任务4 柱塞泵的认知 014
  任务5 液压泵的选用 017
  思考与练习 018

项目2 液压执行元件 019
  任务1 液压马达的认知 019
  任务2 液压缸的认知 024
  思考与练习 032

项目3 液压控制阀 033
  任务1 方向控制阀的认知 033
  任务2 压力控制阀的认知 043
  任务3 流量控制阀的认知 052
  任务4 叠加阀的认知 054
  思考与练习 056

项目4 基本液压回路的安装与调试 058
  任务1 方向控制回路的安装与调试 058
  任务2 压力控制回路的安装与调试 060
  任务3 速度控制回路的安装与调试 064
  任务4 叠加阀控制回路的安装与调试 069
  任务5 比例阀控制回路的安装与调试 073
  思考与练习 076

项目 5 气压传动基础 ·· 077
  任务 1 气压传动系统的工作原理及组成 ·· 077
  任务 2 气压传动系统的特点 ·· 078
  任务 3 气压传动系统的应用 ·· 080
  思考与练习 ··· 081

项目 6 气动元件的认知 ·· 082
  任务 1 气源装置 ··· 082
  任务 2 气动执行元件的认知 ·· 086
  任务 3 气动控制元件的认知 ·· 090
  思考与练习 ··· 096

项目 7 气动回路 ·· 097
  任务 气动基本回路的安装与调试 ··· 097
  思考与练习 ··· 102

项目 1
液压动力元件的认知

任务 1 液压动力元件概述

任务描述

液压系统的动力源来自液压泵。本任务的重点是了解液压泵的特点及工作原理。通过学习,要求掌握液压泵的一些具体参数。

相关知识

一、液压泵的工作原理及特点

液压动力元件为整个液压系统提供动力源,是液压系统的核心元件。最常用的液压动力元件是液压泵。液压泵将原动机(一般是电动机或内燃机)输出的机械能转换为工作液体的压力能,是一种将机械能转化为液压能的能量转换装置。

液压泵由原动机驱动,把输入的机械能转换为油液的压力能,再以压力、流量的形式输入到系统中去。它是液压传动的心脏,也是液压系统的动力源。

1. 液压泵的工作原理

液压泵一般都是靠密封容积变化而工作的,一般称为容积式液压泵,图 1-1-1 所示为单柱塞液压泵的工作原理,图中柱塞 2 装在缸体 3 中形成一个密封容积 a,柱塞在弹簧 4 的作用下一直压紧在偏心轮 1 上。原动机带动着偏心轮 1 旋转,使柱塞 2 做往复运动,从而使密封容积 a 的大小发生周期性的交替变化。当容积 a 由小变大时就形成部分真空,油箱中的油液在大气压作用下,把油经过吸油管顶开单向阀 6 进入容积 a 而实现吸油;反之,当容积 a 由大变小时,a 腔中吸满的油液将顶开单向阀 5 流入液压系统中而实现压油。这样液压泵就不断地实现吸油和压油。

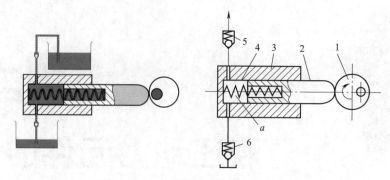

图 1-1-1　单柱塞液压泵的工作原理
1—偏心轮；2—柱塞；3—缸体；4—弹簧；5、6—单向阀

2. 液压泵的特点

单柱塞液压泵具有一切容积式液压泵的基本特点，具体如下：

(1) 具有若干密封且可以周期性变化的空间。液压泵输出流量与此空间的容积变化量和单位时间内的变化次数成正比，与其他因素无关。这是容积式液压泵的一个重要特性。

(2) 油箱内液体的绝对压力必须恒等于或大于大气压力。这是容积式液压泵能够吸入油液的外部条件。因此，为保证液压泵正常吸油，油箱必须与大气相通，或采用密闭的充压油箱。

(3) 具有相应的配流机构，将吸油腔和排液腔隔开，保证液压泵有规律地、连续地吸、排液体。液压泵的结构原理不同，其配油机构也不相同。图 1-1-1 中的单向阀 5、6 就是配油机构。

容积式液压泵中的油腔处于吸油时称为吸油腔。吸油腔的压力决定于吸油高度和吸油管路的阻力，吸油高度过高或吸油管路阻力太大，会使吸油腔真空度过高而影响液压泵的自吸能力；油腔处于压油时称为压油腔，压油腔的压力则取决于外负载和排油管路的压力损失，从理论上来讲排油压力与液压泵的流量无关。

容积式液压泵排油的理论流量取决于液压泵的有关几何尺寸和转速，而与排油压力无关。但排油压力会影响泵的内泄漏和油液的压缩量，从而影响泵的实际输出流量，所以液压泵的实际输出流量随排油压力的升高而降低。

液压泵按其在单位时间内所能输出的油液的体积是否可调节分为定量泵和变量泵两类；按结构形式可分为齿轮式、叶片式和柱塞式三大类。液压泵的图形符号如图 1-1-2 所示。

(a) 单向定量泵　　(b) 单向变量泵　　(c) 双向变量泵

图 1-1-2　液压泵的图形符号

二、液压泵的主要性能参数

1. 压力

(1) 工作压力。液压泵实际工作时的输出压力称为工作压力。工作压力的大小取决于外负载的大小和排油管路上的压力损失,而与液压泵的流量无关。

(2) 额定压力。液压泵在正常工作条件下,按试验标准规定连续运转的最高压力称为液压泵的额定压力。

(3) 最高允许压力。在超过额定压力的条件下,根据试验标准规定,允许液压泵短暂运行的最高压力值,称为液压泵的最高允许压力。

2. 排量和流量

(1) 排量 V。液压泵每转一周,由其密封容积几何尺寸变化计算而得的排出液体的体积叫液压泵的排量。排量可调节的液压泵称为变量泵;排量为常数的液压泵则称为定量泵。

(2) 理论流量 q_i。理论流量是指在不考虑液压泵的泄漏流量的情况下,在单位时间内所排出的液体体积的平均值。显然,如果液压泵的排量为 V,其主轴转速为 n,则该液压泵的理论流量 q_i 为

$$q_i = Vn \tag{1-1}$$

(3) 实际流量 q。液压泵在某一具体工况下,单位时间内所排出的液体体积称为实际流量,它等于理论流量 q_i 减去泄漏流量 Δq,即

$$q = q_i - \Delta q \tag{1-2}$$

(4) 额定流量 q_n。液压泵在正常工作条件下,按试验标准规定(如在额定压力和额定转速下)必须保证的流量。

3. 功率和效率

(1) 液压泵的功率损失。液压泵的功率损失有容积损失和机械损失两部分。

① 容积损失。容积损失是指液压泵流量上的损失,液压泵的实际输出流量总是小于其理论流量,其主要原因是由于液压泵内部高压腔的泄漏、油液的压缩以及在吸油过程中由于吸油阻力太大、油液黏度大及液压泵转速高等原因而导致油液不能全部充满密封工作腔。液压泵的容积损失用容积效率 η_v 来表示,它等于液压泵的实际输出流量 q 与其理论流量 q_i 之比,即

$$\eta_v = \frac{q}{q_i} = \frac{q_i - \Delta q}{q_i} = 1 - \frac{\Delta q}{q_i} \tag{1-3}$$

因此,液压泵的实际输出流量 q 为

$$q = q_i \eta_v = Vn\eta_v \tag{1-4}$$

式中,V 为液压泵的排量(m^3/r);n 为液压泵的转速(r/s)。

液压泵的容积效率随着液压泵工作压力的增大而减小,且随液压泵的结构类型不同而异,但恒小于1。

② 机械损失。机械损失是指液压泵在转矩上的损失。液压泵的实际输入转矩 T_0 总是大

于理论上所需要的转矩 T_i,其主要原因是由于液压泵体内相对运动部件之间因机械摩擦而引起的摩擦转矩损失以及液体的黏性而引起的摩擦损失。液压泵的机械损失用机械效率表示,它等于液压泵的理论转矩 T_i 与实际输入转矩 T_0 之比,设转矩损失为 ΔT,则液压泵的机械效率为

$$\eta_m = \frac{T_i}{T_0} = \frac{1}{1+\frac{\Delta T}{T_i}} \tag{1-5}$$

(2)液压泵的功率。

①输入功率 P_i。液压泵的输入功率是指作用在液压泵主轴上的机械功率,当输入转矩为 T_0,角速度为 ω 时,有

$$P_i = T_0 \omega \tag{1-6}$$

②输出功率 P_o。液压泵的输出功率是指液压泵在工作过程中的实际吸、压油口间的压差 Δp 和输出流量 q 的乘积,即

$$P_o = \Delta p \cdot q \tag{1-7}$$

式中,Δp 为液压泵吸、压油口之间的压力差(N/m^2);q 为液压泵的实际输出流量(m^3/s);P_o 为液压泵的输出功率($N \cdot m/s$ 或 W)。

在实际的计算中,若油箱通大气,液压泵吸、压油的压力差往往用液压泵出口压力 p 代入。

(3)液压泵的总效率。液压泵的总效率是指液压泵的实际输出功率与其输入功率的比值,即

$$\eta = \frac{P}{P_i} = \frac{\Delta p \cdot q}{T_0 \omega} = \frac{\Delta p \cdot q_i \cdot \mu_v}{\frac{T_i \omega}{\eta_m}} = \eta_v \eta_m \tag{1-8}$$

式中,$\Delta p q_i / \omega$ 为理论输入转矩 T_i。

由式(1-8)可知,液压泵的总效率等于其容积效率与机械效率的乘积,所以液压泵的输入功率也可写成

$$P_i = \frac{\Delta p \cdot q}{\eta} \tag{1-9}$$

任务实施

1. 通过网络查询液压泵的主要参数,并列出表格。
2. 分组讨论。每3~5人分为一组,每人在组内简述容积式液压泵的工作原理。
3. 查阅当前我国液压泵发展状况的资料,并客观评价,提出自己的看法。

任务 2　齿轮泵的认知

任务描述

齿轮泵的工作原理是什么？它有什么优缺点？齿轮泵在使用过程中会出现什么问题？齿轮泵主要应用于哪些场合？

相关知识

图 1-2-1 所示为齿轮泵。它看似水泵，但实际功能就像一台抽水机。它能将油液从油箱吸出来。齿轮泵是液压系统中广泛采用的一种液压泵，其主要特点是结构简单、制造方便、价格低廉、体积小、质量小、自吸性能好、对油液污染不敏感、工作可靠；其主要缺点是流量和压力脉动大、噪声大、排量不可调。

图 1-2-1　齿轮泵

一、齿轮泵的工作原理和结构

在齿轮泵的泵体内有一对外啮合齿轮，齿轮两侧有端盖。泵体、端盖和齿轮的各个齿间组成了许多密封工作腔。当齿轮按图 1-2-2 所示方向旋转时，右侧吸油腔由于相互啮合的轮齿逐渐脱开，密封工作腔容积增大，形成部分真空，油箱中的油液被大气压从油箱压入，将齿间槽充满，并随着齿轮旋转，把油液带到左侧压油腔去。在压油区一侧，由于轮齿逐渐进入啮合，密封工作腔容积不断减小，油液便被挤出去。

齿轮泵的原理如图 1-2-2 所示，结构如图 1-2-3 所示。泵的前后盖和泵体由两个定位销定位，用六只螺钉固紧。为了保证齿轮能灵活地转动，同时又要保证泄漏最小，在齿轮端面和泵盖之间应有适当间隙（轴向间隙），对小流量泵轴向间隙为 0.025～0.04 mm，大流量泵为 0.04～0.06 mm。齿顶和泵体内表面间的间隙（径向间隙），由于密封带长，同时齿顶线速度形成的剪切流动又和油液泄漏方向相反，故对泄漏的影响较小。这里要考虑的问题是：当齿轮受到不平衡的径向力后，应避免齿顶和泵体内壁相碰。所以，径向间隙可稍大，一般取 0.13～0.16 mm。为了防止压力油从泵体和泵盖间泄漏到泵外，并减小压紧螺钉的拉力，在泵体两侧的端面上开有油封泄油槽 16，使渗入泵体和泵盖间的压力油引入吸油腔。在泵盖和从动轴上的小孔，用来将泄漏到轴

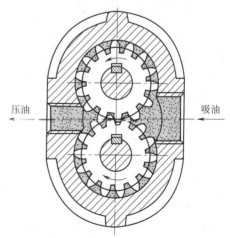

图 1-2-2　外啮合型齿轮泵工作原理

承端部的压力油也引到泵的吸油腔去,防止油液外溢,润滑滚针轴承。

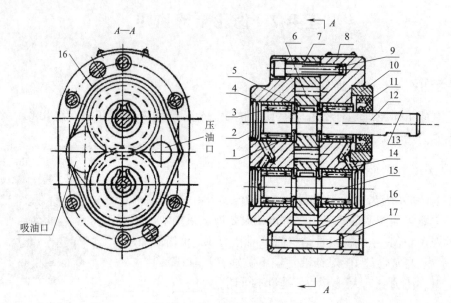

图 1-2-3 齿轮泵的结构

1—轴承外环;2—堵头;3—滚子;4—后泵盖;5—键;6—齿轮;7—泵体;8—前泵盖;9—螺钉;
10—压环;11—密封环;12—主动轴;13—键;14—泄油孔;15—从动轴;16—泄油槽;17—定位销

二、齿轮泵结构特点

1. 泄漏

这里所说的泄漏是指液压泵的内部泄漏,即一部分液压油从压油腔流回吸油腔,没有输送到系统中去。泄漏降低了液压泵的容积效率。外啮合齿轮泵存在三个可能产生泄漏的部位:齿轮端面和端盖间、齿轮外圆和壳体内孔间以及两个齿轮的齿面啮合处。其中对泄漏影响最大的是齿轮端面和端盖间的轴向间隙,这部分泄漏量占总泄漏量的75%~80%,因为这里泄漏途径短,泄漏面积大。轴向间隙过大,泄漏量多,会使容积效率降低;但间隙过小,齿轮端面和端盖间的机械摩擦损失增加,会使泵的机械效率降低。因此,设计和制造时必须严格控制泵的轴向间隙。

2. 齿轮泵的困油问题

为了使齿轮平稳地啮合运转,根据齿轮啮合原理,齿轮的重叠系数应该大于1,即存在两对轮齿同时进入啮合的时候。因此,就有一部分油液困在两对轮齿所形成的封闭容腔之内,如图1-2-4所示。这个封闭容腔先随齿轮转动逐渐减小以后又逐渐增大。减小时会使被困油液受挤压而产生高压,并从缝隙中流出,导致油液发热,同时也使轴承受到不平衡负载的作用;封闭容腔的增大会造成局部真空,使溶于油液中的气体分离出来,产生气穴,这就是齿轮泵的困油现象。其封闭容积的变化如图1-2-4所示。困油现象使齿轮泵产生强烈的噪声和气蚀,这会影响工作平稳性、缩短泵的使用寿命。为了消除困油现象,要在齿轮泵的泵盖上铣出两个困油卸荷凹槽。

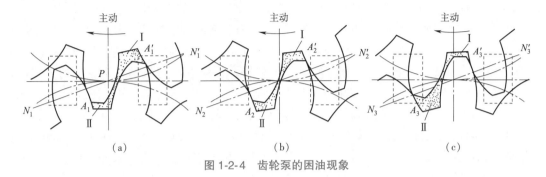

图 1-2-4　齿轮泵的困油现象

3. 径向不平衡力

齿轮泵工作时,在齿轮和轴承上承受径向液压力的作用。如图 1-2-2 所示,泵的右侧为吸油腔,左侧为压油腔。在压油腔内有液压力作用于齿轮上,沿着齿顶的泄漏油,具有大小不等的压力,就是齿轮和轴承受到的径向不平衡力。液压力越高,这个不平衡力就越大,其结果不仅加速了轴承的磨损,降低了轴承的寿命,甚至使轴变形,造成齿顶和泵体内壁的摩擦等。为了解决径向力不平衡问题,在有些齿轮泵上,采用开压力平衡槽的办法来消除径向不平衡力,但这将使泄漏增大、容积效率降低等。CB-B 型齿轮泵则采用缩小压油腔,以减少液压力对齿顶部分的作用面积来减小径向不平衡力,所以泵的压油口孔径比吸油口孔径要小。另外,也可以通过增大径向间隙,使泵体与齿顶有一定的间隙,从而可以减小径向不平衡力。

三、内啮合齿轮泵

内啮合齿轮泵的工作原理也是利用齿间密封容积的变化来实现吸油压油的。图 1-2-5 所示为内啮合齿轮泵的实物以及工作原理图。

 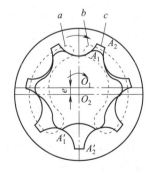

图 1-2-5　内啮合齿轮泵的实物以及工作原理图

它由配油盘(前、后盖)、外转子(从动轮)和偏心安置在泵体内的内转子(主动轮)等组成。内、外转子相差一齿,图 1-2-5 中内转子为六齿,外转子为七齿,由于内外转子是多齿啮合,这就形成了若干密封容积。当内转子围绕中心 O_1 旋转时,带动外转子绕外转子中心 O_2 做同向旋转。这时,由内转子齿顶 A_1 和外转子齿谷 A_2 间形成的密封容积。随着转子的转动密封容积逐渐扩大,于是就形成局部真空,油液从配油窗口 b 被吸入密封腔,至 A_1'、A_2' 位置时封闭容积最大,这时吸油完毕。当转子继续旋转时,充满油液的密封容积便逐渐减小,油液受挤压,于是通过另一配油窗口 a 将油排出。到内转子的另一齿与外转子的齿凹 A_2 全部啮合时,压油完毕。内转子每转一周,由内转子齿顶和外转子齿谷所构成的每个密封容积,完成吸、压油各一次,当内转子连续转动时,即完成了液压泵的吸排油工作。内啮合齿轮泵的外转子齿形是圆弧,内转子齿形为短幅外摆线的等距线,故又称内啮合摆线齿轮泵,也叫转子泵。

内啮合齿轮泵有许多优点,如结构紧凑,体积小,零件少,转速可高达 10 000 r/min,运动平稳,噪声小,容积效率较高等。缺点是流量脉动大,转子的制造工艺复杂等,目前已采用粉末冶金压制成型。随着工业技术的发展,摆线齿轮泵的应用将会愈来愈广泛。由于内啮合齿轮泵可以正转也可以反转,因此可以用作液压马达。

任务实施

1. 讲述齿轮泵的结构,并说出其主要由哪几部分组成。
2. 分组讨论。每 3~5 人分为一组,每人在组内简单的讲解齿轮泵的工作原理。
3. 写出齿轮泵的优缺点,并说出其应用场合。

任务 3 叶片泵的认知

任务描述

某液压设备需要配备一个液压泵,要求具有结构紧凑、流量均匀、噪声小、运转平稳等特点。同时还要求,这个液压泵为中压或者低压泵,工作环境比较清洁。请查阅资料,为这台设备选择一个适合的液压泵,并分析其工作原理、结构特点和优缺点。

相关知识

一、单作用叶片泵

1. 单作用叶片泵的工作原理

单作用叶片泵实物如图 1-3-1 所示,工作原理如图 1-3-2 所示。单作用叶片泵由转子 1、定子 2、叶片 3 和端盖等组成。定子具有圆柱形内表面,定子和转子间有偏心距。叶片装在转子槽中,并可在槽内滑动,当转子回转时,由于离心力的作用,使叶片紧靠在定子内壁,这样在定子、转子、叶片和两侧配油盘间就形成若干密封的工作空间。当转子按图示的方向回转时,在图的右部,叶片逐渐伸出,叶片间的工作空间逐渐增大,从吸油口吸油,这是吸油腔。在图的左部,叶片被定子内壁逐渐压进槽内,工作空间逐渐缩小,将油液从压油口压出,这是压油腔,在吸油腔和压油腔之间,有一段封油区,把吸油腔和压油腔隔开,这种叶片泵在转子每转一周,每个工作空间完成一次吸油和压油,因此称为单作用叶片泵。转子不停地旋转,泵就不断地吸油和排油。

图 1-3-1　单作用叶片泵实物

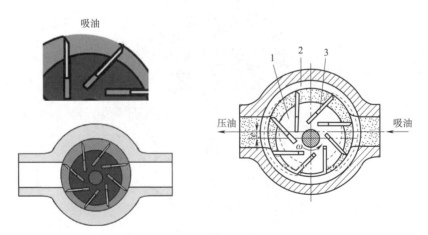

图 1-3-2　单作用叶片泵的工作原理

1—转子;2—定子;3—叶片

2. 单作用叶片泵的结构特点

(1)改变定子和转子之间的偏心距可改变流量。

(2)处在压油腔的叶片顶部受到压力油的作用,将叶片推入转子槽内。为了使叶片顶部可靠地和定子内表面相接触,压油腔一侧的叶片底部要通过一个特殊的沟槽和压油腔相通。吸油腔一侧的叶片底部要和吸油腔相通,这里的叶片仅靠离心力的作用顶在定子内表面上。

(3)由于转子受到不平衡的径向液压作用力,所以这种泵一般不宜用于高压泵。

(4)为了更有利于叶片在惯性力作用下向外伸出,而使叶片有一个与旋转方向相反的倾斜角,称后倾角,一般为24°。

二、双作用叶片泵

1. 双作用叶片泵的工作原理

双作用叶片泵的工作原理如图1-3-3所示,泵也是由定子1、转子2、叶片3和配油盘等组成。转子中心和定子中心重合,定子内表面近似为椭圆柱形,该椭圆柱形由两段长半径R、两段短半径r和四段过渡曲线所组成。当转子转动时,叶片在离心力的作用下,在转子槽内做径向移动而压向定子内表,由叶片、定子的内表面、转子的外表面和两侧配油盘间形成若干密封空间。当转子按图示方向旋转时,处在小圆弧上的密封空间经过渡曲线而运动到大圆弧的过程中,叶片外伸,密封空间的容积增大,要吸入油液;再从大圆弧经过渡曲线运动到小圆弧的过程中,叶片被定子内壁逐渐压进槽内,密封空间容积变小,将油液从压油口压出。显然,当转子每转一周,每个工作空间要完成两次吸油和压油,所以称之为双作用叶片泵。这种叶片泵由于有两个吸油腔和两个压油腔,并且各自的中心夹角是对称的,所以作用在转子上的油液压力平衡。为了使径向力完全平衡,密封空间数的叶片数应当是双数。

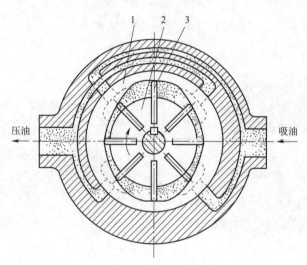

图1-3-3 双作用叶片泵的工作原理
1—定子;2—转子;3—叶片

2. 双作用叶片泵的结构特点

(1)配油盘。

双作用叶片泵的配油盘如图1-3-4所示。配油盘是泵的配油机构。为了保证配油盘的吸、压油窗口在工作中能隔开,就必须使配油盘上封油区夹角 α 大于或等于两个相邻叶片间的夹角。此外,还要求定子圆弧部分的夹角 β≥α,以免产生困油和气穴现象。从图 1-3-4 可以看出,在配油盘的压油窗口上开有一个三角槽,它的作用主要是用来减小泵的流量脉动和压力脉动。封油区中两相邻叶片之间的油液其压力基本与吸油区压力相同,当这部分液体从封油区到达压油窗口时,相当于一个低压区域突然和一个高压区域接通,这势必造成压油腔中的油液倒流进来,引起泵输出流量和压力的脉动。在配油盘上叶片从封油区进入压油窗口的一边开三角槽,可使那块低压液体逐渐进入压油窗口,压力逐渐上升,从而降低泵的流量脉动和压力脉动。三角槽的尺寸通常由实验来确定。

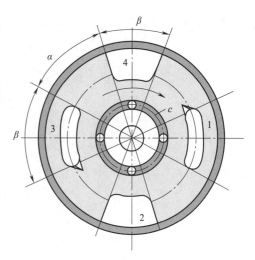

图 1-3-4　双作用叶片泵的配油盘

(2)定子曲线。

定子的过渡曲线如图 1-3-5 所示,它由四段圆弧和四段过渡曲线组成。理想的过渡曲线应保证使叶片在转子槽中滑动时径向速度和加速度变化均匀,保证叶片对定子表面的冲击尽可能小。

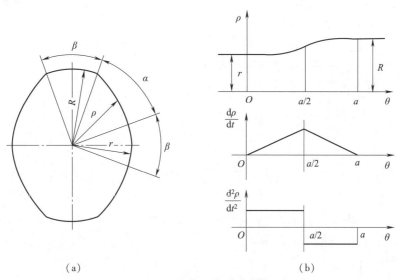

图 1-3-5　定子的过渡曲线

(3)叶片的倾角。

叶片在转子中的安放应当有利于叶片的滑动,磨损要小。图 1-3-6 给出了双作用叶片泵的倾角。在工作过程中,受离心力和叶片根部压力油的作用,叶片紧紧地与定子接触。定子内表面给叶片顶部的反作用力 N 可分解为两个力,即与叶片垂直的力 T 和沿叶片槽方向的力 P。显然,力 T 容易使叶片折断。为此,通常将转子槽按旋转方向倾斜 α 角,这样可以减小力 T 的值。由理论分析和实验验证,一般取 α 为 $10°\sim14°$。

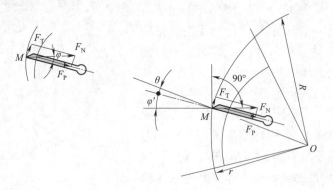

图 1-3-6　双作用叶片泵的倾角

3. 提高双作用叶片泵压力的措施

由于一般双作用叶片泵的叶片底部有压力油,使得处于吸油区的叶片顶部和底部的液压作用力不平衡。叶片顶部以很大的压紧力抵在定子吸油区的内表面上,使磨损加剧,影响叶片泵的试用寿命,尤其是工作压力较高时,磨损更严重。因此,吸油区叶片两端压力不平衡,限制了双作用叶片泵工作压力的提高。所以,在高压叶片泵的结构上必须采取措施使叶片压向定子的作用力减小。常用的措施有:

(1)减小作用在叶片底部的油液压力。
(2)减小叶片底部承受压力油作用的面积。
(3)使叶片顶部和底部的液压作用力平衡。

三、双级叶片泵和双联叶片泵

1. 双级叶片泵

为了得到较高的工作压力,也可以不用高压叶片泵,而用双级叶片泵。双级叶片泵是由两个通压力的单级叶片泵装在一个泵体内在油路上串接而成的。双级叶片泵的工作原理如图 1-3-7 所示,两个单级叶片泵的转子装在同一根传动轴上,当传动轴回转时就带动两个转子一起转动,就可以产生双倍的工作压力。

2. 双联叶片泵

双联叶片泵是由两个单级叶片泵装在一个泵体内在油路上并联组成的。两个叶片泵的转子由同一传动轴带动旋转,并各有独立的出油口,两个泵可以是相等流量的,也可以是不等流量的。

双联叶片泵常用于有快进和工作进给要求的机械加工专用机床中。在这种机床中,双联泵由一小个流量泵和一个大流量泵组成。当快速进给时,两个泵同时供油(此时压力较低);

当工作进给时,由小流量泵供油(此时压力较高),同时在油路系统上使大流量泵卸荷,这与采用一个高压大流量的泵相比,可以节省能源,减少油液发热。这种双联叶片泵也常用于机床液压系统中需要两个互不影响的独立油路中。

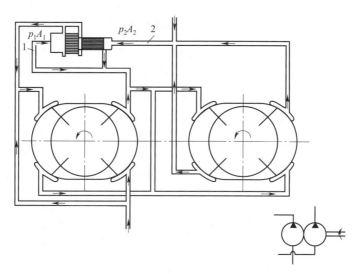

图 1-3-7　双级叶片泵的工作原理
1、2—管路

3. 叶片泵的主要性能

叶片泵的结构较齿轮泵复杂,但其工作压力较高,且流量脉动小,工作平稳,噪声较小,寿命较长。所以它被广泛应用于机械制造中的专用机床、自动线等中低液压系统中,但其结构复杂,吸油特性不太好,对油液的污染也比较敏感。叶片泵的主要性能参数如下:

(1)压力:中低压叶片泵的额定压力一般为 6.3 MPa,双作用高压叶片泵的最高工作压力可达 28~30 MPa,变量叶片泵的压力一般不超过 17.5 MPa。

(2)排量:叶片泵的排量范围为 0.5~4 200 mL/r,常用的双作用叶片泵为 2.5~300 mL/r,变量叶片泵为 6~120 mL/r。

(3)转速:小排量双作用叶片泵的最高转速可达 8 000~10 000 r/min,一般排量的叶片泵为 1 500~2 000 r/min,常用的变量叶片泵最高转速大约为 3 000 r/min,最低转速为 600~900 r/min。

(4)效率:双作用叶片泵的容积效率较高,可达 93%~95%,但机械效率较低,其总效率与齿轮泵差不多。

(5)寿命:叶片泵的寿命高于齿轮泵,高压叶片泵的使用寿命可达 5 000 h 以上。

任务实施

1. 通过本任务的学习,为液压设备平台选择一个合适的叶片泵。
2. 分组讨论。每 3~5 人分为一组,列出叶片泵和齿轮泵的区别。

3. 查阅资料,找出还有哪些常用的叶片泵,如限量式叶片泵。分析限量式叶片泵的工作原理。

任务4 柱塞泵的认知

任务描述

某型号龙门刨床需要配备一个液压泵,要求该泵具有压力高、流量大、功率大、结构紧凑、占用空间小、效率高、流量调节方便等特点。应该配备一个什么样的液压泵呢?请分析所选泵的结构及特点。

相关知识

一、径向柱塞泵

柱塞泵实物如图1-4-1所示。径向柱塞泵的工作原理如图1-4-2所示。柱塞泵由柱塞、转子(缸体)、衬套、定子和配油轴组成。定子和转子之间有一个偏心e。衬套固定在转子孔内随之一起转动。配油轴是固定不动的。柱塞在转子的径向孔内运动,形成了泵的密封工作容腔。显然,当转子按图示方向转动时,位于上半周的工作容腔处于吸油状态,油箱中的油液经配油轴的a孔进入b腔;位于下半周的工作容腔则处于压油状态,c腔中的油将从配油轴的d孔向外输出。改变定子与转子偏心距e的大小和方向,就可以改变泵的输出流量和泵的吸、压油方向。

图1-4-1 柱塞泵实物

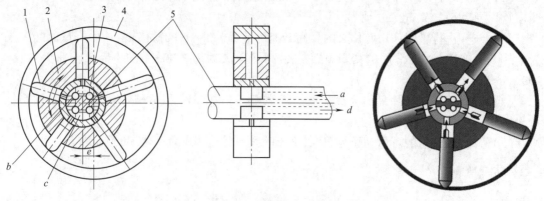

图1-4-2 径向柱塞泵的工作原理

1—柱塞;2—转子;3—衬套;4—定子;5—配油轴

二、轴向柱塞泵

1. 轴向柱塞泵的工作原理

轴向柱塞泵是将多个柱塞配置在一个共同缸体的圆周上,并使柱塞中心线和缸体中心线平行的一种泵。轴向柱塞泵有两种形式:直轴式(斜盘式)和斜轴式(摆缸式)。图 1-4-3 所示为直轴式轴向柱塞泵的工作原理。这种泵的主体由缸体 1、配油盘 2、柱塞 3 和斜盘 4 组成。柱塞沿圆周均匀分布在缸体内;斜盘轴线与缸体轴线倾斜一角度;柱塞靠机械装置或在低压油作用下压紧在斜盘上(图中为弹簧);配油盘 2 和斜盘 4 固定不转。当原动机通过传动轴使缸体转动时,由于斜盘的作用,迫使柱塞在缸体内作往复运动,并通过配油盘的配油窗口进行吸油和压油。如图 1-4-3 中所示回转方向,当缸体转角在 $\pi \sim 2\pi$ 范围内,柱塞向外伸出,柱塞底部缸孔的密封工作容积增大,通过配油盘的吸油窗口吸油;在 $0 \sim \pi$ 范围内,柱塞被斜盘推入缸体,使缸孔容积减小,通过配油盘的压油窗口压油。缸体每转一周,每个柱塞各完成吸、压油一次。轴向柱塞泵具有结构紧凑、工作压力高、容易实现变量等优点。

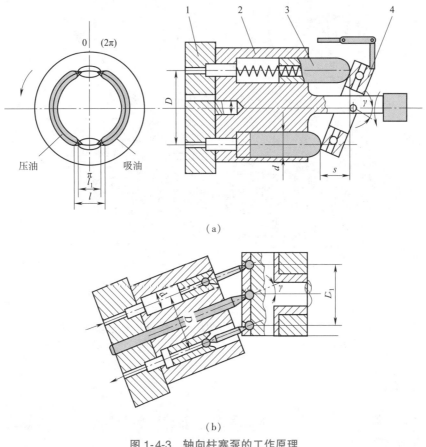

图 1-4-3 轴向柱塞泵的工作原理
1—缸体;2—配油盘;3—柱塞;4—斜盘

斜轴式轴向柱塞泵如图 1-4-3(b)中的缸体轴线相对传动轴轴线成一倾角,传动轴端部用万向铰链、连杆与缸体中的每个柱塞相联结。当传动轴转动时,通过万向铰链、连杆使柱塞和缸体一起转动,并迫使柱塞在缸体中做往复运动,借助配油盘进行吸油和压油。这类泵的优点是变量范围大、泵的强度较高,但和上述直轴式柱塞泵相比,结构较复杂,外形尺寸和质量均较大。

2. 轴向柱塞泵的结构特点

图 1-4-4 所示为一种轴向柱塞泵的结构。传动轴 6 通过花键带动缸体 3 旋转。柱塞 7(七个)均匀安装在缸体上。柱塞的头部装有滑靴 9,滑靴与柱塞是球铰连接,可以任意转动。由弹簧将滑靴压靠在斜盘 11 上。这样,当缸体转动时,柱塞就可以在缸体中往复运动,完成吸油和压油过程。配油盘 4 与泵的吸油口和压油口相通,固定在泵体上。另外,在滑靴与斜盘相接触的部分有一个油室,压力油通过柱塞中间的小孔进入油室,在滑靴与斜盘之间形成一个油膜,起着静压支承作用,从而减少了磨损。

这种泵的变量机构是手动的。转动手轮 15,通过调整丝杠螺母副可以改变斜盘的倾角,从而改变泵的输出流量。

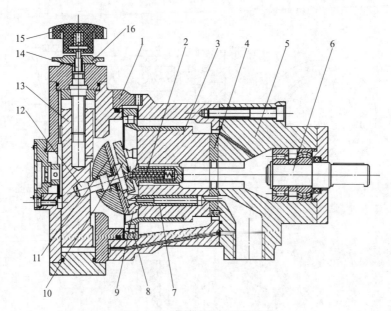

图 1-4-4 直轴式向柱塞泵结构

1—泵体;2—弹簧;3—缸体;4—配油盘;5—前泵体;6—传动轴;7—柱塞;8—轴承;9—滑靴;
10—回程盘;11—斜盘;12—轴销;13—变量活塞;14—丝杠;15—手轮;16—螺母

任务实施

1. 通过本任务的学习,为龙门刨床选择一个合适的柱塞泵,并确定其功率、能够承受的最大压力、流量等参数。

2. 分组讨论。每 3~5 人分为一组,列出叶片泵、齿轮泵、柱塞泵三者之间的差异。

任务 5　液压泵的选用

任务描述

如图 1-5-1 所示,某液压设备平台液压站已装有两个液压泵。一个为图 1-5-2(a)中的变量叶片泵;另外一个是图 1-5-2(b)中的柱塞泵。请查阅设备说明书等相关资料或根据前面所学知识,列出选择这两个泵的理由。

图 1-5-1　某液压设备平台液压站

(a)变量叶片泵　　　　　　　(b)柱塞泵

图 1-5-2　液压泵

相关知识

液压泵是液压系统提供一定流量和压力的油液动力元件,它是每个液压系统不可缺少的核心元件。合理选择液压泵对于降低液压系统能耗、提高系统效率、降低噪声、改善工作性能和保证系统的可靠工作都十分重要。

选择液压泵的原则是：根据主机工况、功率大小和系统对工作性能的要求，首先确定液压泵的类型，然后按系统所要求的压力、流量大小确定其规格型号，还要考虑价格、维护方便与否等问题。表 1-5-1 列出了液压系统中常用液压泵的主要性能和应用场合。

表 1-5-1 液压系统中常用液压泵的性能比较

性能	外啮合齿轮泵	双作用叶片泵	限压式变量叶片泵	径向柱塞泵	轴向柱塞泵	螺杆泵
输出压力	低压	中压	中压	高压	高压	低压
流量调节	不能	不能	能	能	能	不能
效率	低	较高	较高	高	高	较高
输出流量脉动	很大	很小	一般	一般	一般	最小
自吸特性	好	较差	较差	差	差	好
对油的污染敏感性	不敏感	较敏感	较敏感	很敏感	很敏感	不敏感
噪声	大	小	较大	大	大	最小

一般来说，由于各类液压泵各自突出的特点，其结构、功用和动转方式各不相同，因此应根据不同的使用场合选择合适的液压泵。一般在机床液压系统中，往往选用双作用叶片泵和限压式变量叶片泵；而在筑路机械、港口机械以及小型工程机械中，往往选用抗污染能力较强的齿轮泵；在负载大、功率大的场合中，往往选择柱塞泵。

任务实施

1. 通过网络查询液压泵的选用原则，并进行归纳总结。
2. 分组讨论。每 3~5 人分为一组，对叶片泵、齿轮泵、柱塞泵等常用液压泵的差异进行归纳总结。
3. 找出常用液压设备液压站使用的液压泵的型号，如数控车床中的液压夹紧系统。并写出选用依据。

思考与练习

1. 找出某液压站上使用的液压泵的型号，根据铭牌写出其压力流量等参数。
2. 选择液压泵时需要考虑哪些性能？
3. 总结所学液压泵的一些基本特点以及常用场合。

项目 2
液压执行元件

任务 1 液压马达的认知

任务描述

了解常见液压马达的基本工作原理及符号,熟悉液压马达的种类及常见液压马达的不同应用场景。

相关知识

一、液压马达的概念及符号

液压马达是一种把液体的压力能转换为机械能的装置,如图 2-1-1 所示。从结构上讲,同类型的液压泵和液压马达基本上很相似,但由于两者的工作情况不同,使两者在结构上存在些许差异。从原理上讲,液压马达也可以作为液压泵使用,液压泵可以作为液压马达用。两者存在的差别主要有:

(1) 液压泵一般是单方向旋转的,而液压马达一般需要正反转,所以后者在内部结构上具有对称性。

(2) 为减小径向力,减小吸油阻力,一般液压泵的吸油口比出油口的尺寸要大,而液压马达低压腔的压力要稍微高于大气压力。

图 2-1-1 液压马达实物

(3) 液压马达要求能够在很宽的转速范围情况下能够正常工作,因此,应采用液动轴承或静压轴承。因为当马达速度很低时,如果采用动压轴承,不易形成润滑滑膜。

(4) 叶片液压泵依靠叶片和转子一起高速旋转而产生一种离心力,这种离心力能够使叶片始终紧贴定子的内表面,从而起到封油的作用,形成工作容积。若将其当马达使用,则必须在液压马达的叶片根部装上弹簧,以此来保证叶片始终紧贴定子内表面,以便马达能够正常启动。

（5）在结构上,液压泵需要保证其具有自吸能力,而对于液压马达来说没有这一要求。

（6）液压马达必须具有较大的启动扭矩。所谓的启动扭矩,就是指马达由静止状态启动时,马达轴上所能输出的扭矩。该扭矩通常大于在同一工作压差时处于运行状态下的扭矩。所以,为了使启动扭矩尽可能接近工作状态下的扭矩,要求马达扭矩的脉动小,内部摩擦也小。

由于液压泵与液压马达具有上述不同的特性,所以很多类型的液压泵和液压马达不能互通使用。

液压马达根据其额定转速不同,可分为低速液压马达和高速液压马达两大类。额定转速低于 500 r/min 的属于低速液压马达,额定转速高于 500 r/min 的属于高速液压马达。

液压马达也可根据其结构类型的不同,分为齿轮式、叶片式、柱塞式和其他型式。

液压马达的图形符号见表 2-1-1。

表 2-1-1　液压马达的图形符号

定量液压泵-马达	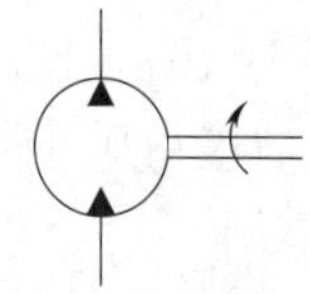	双向定量马达	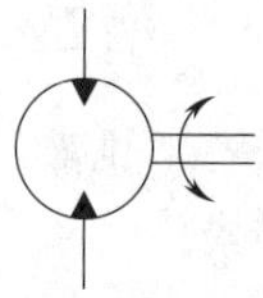
变量液压泵-马达	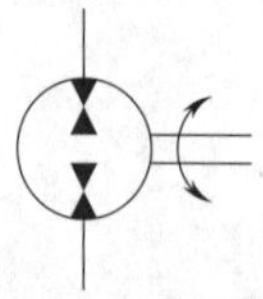	单向变量马达	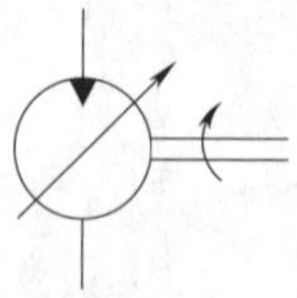
单向定量马达	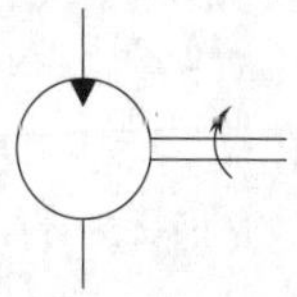	双向变量马达	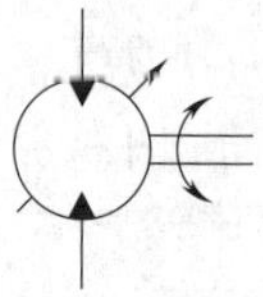

二、液压马达的性能参数

液压马达的性能参数很多。下面是液压马达的主要性能参数:

1. 排量、流量和容积效率

通常将马达的轴每转一周,按几何尺寸计算所进入的液体容积,称之为马达的排量 V,也称作理论排量、几何排量,即不考虑泄漏损失时的排量。

液压马达的排量代表其工作容腔的大小。在工作中,输出的转矩大小由负载转矩决定。因此,排量是一个重要的参数。但是,推动同样大小的负载,工作容腔大的马达的压力要低于工作容腔小的马达的压力。所以,工作容腔的大小是液压马达工作能力的主要指标,也就是说,排量的大小是液压马达工作能力的重要标志。

根据液压动力元件的工作原理可知,马达转速 n、理论流量 q_i 与排量 V 之间具有下列关系

$$q_i = nV \tag{2-1}$$

式中,q_i 为理论流量(m^3/s);n 为转速(r/min);V 为排量(m^3/s)。

为了满足转速要求,通常马达实际输入流量 q 大于理论输入流量,则有

$$q = q_i + \Delta q \tag{2-2}$$

式中,Δq 为泄漏流量。

$$\eta_v = q_i/q = 1/(1+\Delta q/q_i) \tag{2-3}$$

所以得实际流量

$$q = q_i/\eta_v \tag{2-4}$$

2. 液压马达输出的理论转矩

根据排量的大小,在给定压力下可以计算液压马达所能输出的转矩的大小,也可以在给定的负载转矩下计算马达的工作压力的大小。当液压马达进出油口之间的压力差为 ΔP,输入液压马达的流量为 q,液压马达输出的理论转矩为 T_t,角速度为 ω。如若不计损失,液压马达输入的液压功率应当全部转化成液压马达输出的机械功率,即

$$\Delta P \cdot q = T_t \omega \tag{2-5}$$

又因为 $\omega = 2\pi n$,所以液压马达的理论转矩为

$$T_t = \Delta P \cdot V/2\pi \tag{2-6}$$

式中,ΔP 为马达进出口之间的压力差。

3. 液压马达的机械效率

由于液压马达内部存在不可避免的各种摩擦,实际输出的转矩 T 总要比理论转矩 T_t 小一些,即

$$T = T_t \eta_m \tag{2-7}$$

式中,η_m 为液压马达的机械效率(%)。

4. 液压马达的启动机械效率 η_m

液压马达的启动机械效率指的是液压马达由静止状态开始启动时,马达实际输出的转矩 T_0 与它在同一工作压差时的理论转矩 T_t 之比,即

$$\eta_m = T/T_t \tag{2-8}$$

液压马达的启动机械效率同样也是启动性能的指标。因为在同样的压力下,液压马达由静止到开始启动的状态中,其输出转矩要比运转中的转矩大,这给液压马达带载启动造成了困难。所以,启动性能对液压马达非常重要。启动机械效率能够反映其启动性能的高低。

在实际工作中一般希望启动性能好,即希望启动转矩和启动机械效率能够高一些。现将不同结构形式的液压马达的启动机械效率 η_{m0} 的大致数值列入表 2-1-2 中。

表2-1-2　液压马达的启动机械效率

液压马达的结构形式		启动机械效率 η_{m0}/%
齿轮马达	老结构	0.60~0.80
	新结构	0.85~0.88
叶片马达	高速小扭矩型	0.75~0.85
轴向柱塞马达	滑履式	0.80~0.90
	非滑履式	0.82~0.92
曲轴连杆马达	老结构	0.80~0.85
	新结构	0.83~0.90
静压平衡马达	老结构	0.80~0.85
	新结构	0.83~0.90
多作用内曲线马达	由横梁的滑动摩擦副传递切向力	0.90~0.94
	传递切向力的部位具有滚动副	0.95~0.98

由表2-1-2可知，启动性能最好的是多作用内曲线马达，轴向柱塞马达、曲轴连杆马达和静压平衡马达居中，叶片马达较差，而齿轮马达最差。

5. 液压马达的转速

液压马达的转速取决于油液的流量和液压马达本身的排量 V，可用式(2-9)计算：

$$n_t = q_i / V \tag{2-9}$$

式中，n_t 为理论转速(r/min)。

由于液压马达内部有泄漏，有一部分液体因泄漏损失掉了，所以并不是所有进入马达的液体都推动液压马达做功，因此液压马达的实际转速要比理论转速低一些。

$$n = n_t \eta_v \tag{2-10}$$

式中，n 为液压马达的实际转速(r/min)；η_v 为液压马达的容积效率(%)。

6. 最低稳定转速

最低稳定转速是指液压马达在额定负载下，不出现爬行现象的最低转速。所谓爬行现象，指的是当液压马达在工作中转速过低时，往往保持不了均匀的速度，进入时动时停的不稳定状态。

液压马达在低速时产生爬行现象的原因是：

(1) 摩擦力的大小不稳定。

通常的摩擦力是随着速度增大而增加的，而对于静止或者低速区域工作的马达而言，其内部的摩擦阻力在工作速度增大时非但不增加，反而减少，形成了所谓"负特性"的阻力。另一方面，液压马达和负载是由液压油被压缩后压力升高而被推动的。因此，可用图2-1-2(a)所示的物理模型表示低速区域液压马达的工作过程：以匀速 v_0 推弹簧的一端(相当于高压下不可压缩的工作介质)，使质量为 m 的物体(相当于马达和负载质量、转动惯量)克服"负特性"的摩擦阻力而运动。当物体静止或速度很低时阻力大，弹簧不断压缩，从而需要增加推力。只有等到弹簧压缩到其推力大于静摩擦力时,物体才开始运动。一旦物体开始运动,如果阻力突然减小,物体会突然加速移动。结果将使弹簧压缩量减少,推力减小,物体依靠惯性前移一段路程后停止下来。弹簧再次移动,又使弹簧压缩,推力增加,物体就再一次移动和跃动,形成如图 2-1-2(b)所示的时动时停的状态。这种现象称为液压马达的爬行现象。

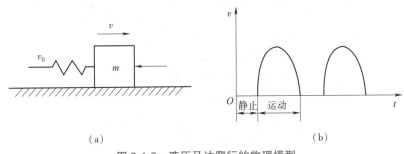

图 2-1-2　液压马达爬行的物理模型

(2)泄漏量大小不稳定。

液压马达的泄漏量随着转子转动的相位角度变化作周期性波动,因此泄漏量不是每个瞬间都是相同的。由于低速时进入马达的流量小,相对而言泄漏所占的比重就较大,泄漏量的不稳定就会明显地影响到参与马达工作的流量数值上,从而造成转速的波动。当马达在低速运转时,其转动部分及所带的负载表现出的惯性较小,上述影响比较明显,因而出现爬行现象。在实际工作中,一般期望最低稳定转速越小越好。

7. 最高使用转速

液压马达的最高使用转速主要受使用寿命和机械效率的限制。转速提高后,各运动副的磨损加剧,从而使用寿命降低。转速高也就意味着液压马达需要输入的流量大,造成各过流部分的流速相应增大,压力损失也随之增加,从而使机械效率降低。

对于某些液压马达而言,转速的提高还受到背压的限制。例如曲轴连杆式液压马达,转速提高时,回油背压必须显著增大才能保证连杆不会撞击曲轴表面,避免出现撞击现象。随着转速的提高,回油腔所需的背压值也应随之提高。但过度提高背压,会使液压马达的效率明显下降。为了使液压马达的效率不致过低,其转速不应太高。

8. 调速范围

液压马达的调速范围用最高使用转速和最低稳定转速之比表示,即

$$i = n_{max}/n_{min} \tag{2-11}$$

三、液压马达的工作原理

常用的液压马达的结构与同类型的液压泵相似。下面以叶片马达的工作原理为例进行介绍。

图 2-1-3 为叶片液压马达工作原理。

当压力为 p 的油液从进油口进入叶片 1 和 3 之间时,叶片 2 因两面均受液压油的作用,所

以不产生转矩。叶片 1、3 上，一面作用有压力油，另一面为低压油，但由于叶片 3 伸出的面积大于叶片 1 伸出的面积，因此作用于叶片 3 上的总液压力要比作用于叶片 1 上的总液压力大，于是压力差使得转子产生了顺时针转矩。同样道理，压力油进入叶片 5 和 7 之间时，叶片 7 伸出的面积大于叶片 5 伸出的面积，同样也产生顺时针转矩。这样就把油液的压力能转变成了机械能，这就是叶片马达的工作原理，而当输油方向改变时，液压马达就反转。

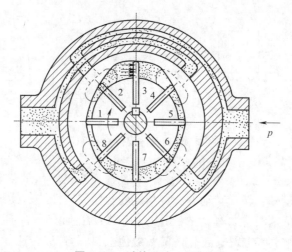

图 2-1-3　叶片马达工作原理

当定子的长短径差值越大，转子的直径越大，以及输入的压力越高时，叶片马达输出的转矩也就会越大。

叶片马达的体积小，使得转动惯性小，动作比较灵敏，换向频率适应性高。但其泄漏较大，不能够在很低的转速下工作，故叶片马达一般适用于转速高、转矩小和动作灵敏的场合。

任务实施

1. 画出设备上的液压马达的图形符号。
2. 能熟练调节马达的转速。
3. 分组讨论，总结出各种液压马达对应的应用场景。

任务 2　液压缸的认知

任务描述

了解常见液压缸的基本工作原理及符号，熟悉常见液压缸的种类及常见液压缸的不同应用场景。

相关知识

一、液压缸的类型特点及符号

液压缸又称油缸。它是液压系统中的一种执行元件,其功能是将液压能转变成机械能。液压缸的种类很多,其详细分类及符号见表2-2-1。

表2-2-1 常见液压缸的种类特点及符号

分类	名称	符号	说明
单作用液压缸	柱塞式液压缸	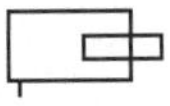	柱塞仅单向运动,返回行程是利用自重或负荷将柱塞推回
	单活塞杆液压缸	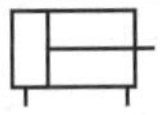	活塞仅单向运动,返回行程是利用自重或负荷将活塞推回
	双活塞杆液压缸	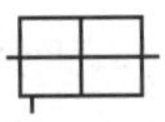	活塞的两侧都装有活塞杆,只能向活塞一侧供给压力油,返回行程通常利用弹簧力、重力或外力
	伸缩液压缸	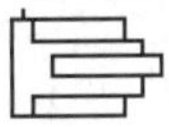	它以短缸获得长行程。用液压油由大到小逐节推出,靠外力由小到大逐节缩回
双作用液压缸	单活塞杆液压缸	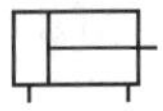	单边有杆,两向液压驱动,两向推力和速度不等
	双活塞杆液压缸	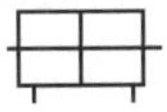	双向有杆,双向液压驱动,可实现等速往复运动
	伸缩液压缸	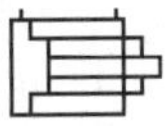	双向液压驱动,伸出由大到小逐步推出,由小到大逐节缩回
组合液压缸	弹簧复位液压缸	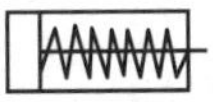	单向液压驱动,由弹簧力复位
	串联液压缸	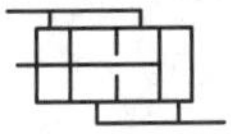	用于缸的直径受限制,而长度不受限制处,获得大的推力
	增压缸(增压器)	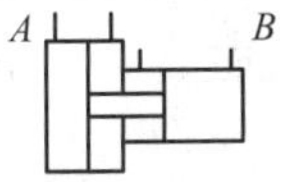	由低压力室 A 缸驱动,使 B 室获得高压油源
	齿条传动液压缸		活塞往复运动经装在一起的齿条驱动齿轮获得往复回转运动
摆动液压缸	摆动马达		输出轴直接输出扭矩,其往复回转的角度小于360°,也称摆动马达

二、常用的液压缸

1. 活塞式液压缸

活塞式液压缸根据其使用要求的不同,分为双杆式和单杆式两种。

(1)双杆式活塞缸如图2-2-1所示。活塞两端都有一根直径相等的活塞杆伸出的液压缸称为双杆式活塞缸,它一般由缸盖、活塞、缸体、活塞杆和密封件等零件构成。根据安装方式的不同,可分为缸筒固定式和活塞杆固定式两种。

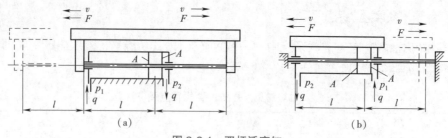

图 2-2-1　双杆活塞缸

图 2-2-1(a)所示为缸筒固定式的双杆活塞缸。它的进出油口于缸筒两端,活塞通过活塞杆带动工作台移动。当活塞的有效行程为 l 时,整个工作台的运动范围为 $3l$。所以,其机床占地面积比较大,一般不适合用于小型机床。当对工作台行程的要求较高时,可采用图 2-2-1(b)所示的活塞杆固定形式,此时的缸体与工作台相连,通过支架,活塞杆固定在机床上,动力由缸体传出。对于这种安装形式,工作台的移动范围只等于液压缸有效行程 l 的两倍($2l$),因此占地比较面积小。然而,其进出油口可以设置在固定不动的空心的活塞杆的两端,但它们必须使用软管连接。

由于双杆活塞缸两端的活塞杆直径通常是相等的,因此它左、右两腔的有效面积也相等。当分别向左、右腔输入相同压力和相同流量的油液时,液压缸左、右两个方向的推力和速度也是一样的。当活塞的直径为 D,活塞杆的直径为 d,液压缸进、出油腔的压力为 p_1 和 p_2,输入流量为 q 时,双杆活塞缸的推力 F 和速度 v 为

$$F=A(p_1-p_2)=\pi(D^2-d^2)(p_1-p_2)/4 \tag{2-12}$$

$$v=q/A=4q/\pi(D^2-d^2) \tag{2-13}$$

式中,A 为活塞的有效工作面积。

双杆活塞缸在工作时,所设计的活塞杆一个是受拉的,而另一个是不受力的。因此,这种液压缸的活塞杆可以做得细一些。

(2)单杆式活塞缸。如图 2-2-2 所示,单杆液压缸分为缸体固定和活塞杆固定两种形式,其活塞只有一端带活塞杆,但它们的工作台移动范围都是活塞有效行程的两倍。

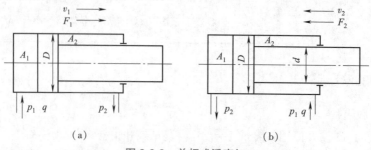

图 2-2-2　单杆式活塞缸

由于液压缸两腔的有效工作面积不等,因此它在两个方向上的输出推力和速度也不等,其值分别为

$$F_1 = (p_1A_1 - p_2A_2) = \pi[(p_1-p_2)D^2 - p_2d^2]/4 \quad (2\text{-}14)$$

$$F_1 = (p_1A_1 - p_2A_2) = \pi[(p_1-p_2)D^2 - p_2d^2]/4 \quad (2\text{-}15)$$

$$v_1 = q/A_1 = 4q/\pi D^2 \quad (2\text{-}16)$$

$$v_2 = q/A_2 = 4q/\pi(D^2 - d^2) \quad (2\text{-}17)$$

由式(2-14)~式(2-17)可知,由于 $A_1 > A_2$,所以 $F_1 > F_2$, $v_1 < v_2$。如把两个方向上的输出速度 v_2 和 v_1 的比值称为速度比,记作 λ_v,则 $\lambda_v = v_2/v_1 = 1/[1-(d/D)^2]$。因此,$d = D\sqrt{(\lambda_v - 1)/\lambda_v}$。在已知 D 和 λ_v 时,可确定 d 值。

差动液压缸。单杆活塞缸在其左右两腔都接通高压油时称为"差动连接",如图 2-2-3 所示。差动连接缸左右两腔的油液压力相同,但由于左腔(无杆腔)的有效面积大于右腔(有杆腔)的有效面积,故活塞向右运动的同时,使右腔中排出的油液(流量为 q')也进入左腔,加大了流入左腔的流量($q+q'$),从而加快了活塞移动的速度。实际上活塞在运动时,由于差动连接时两腔间的管路中有压力损失,所以右腔中油液的压力稍大于左腔油液压力。而这个差值一般都较小,可以忽略不计。因此,差动连接时活塞推力 F_3 和运动速度 v_3 为

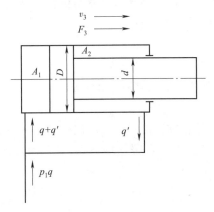

图 2-2-3 差动缸

$$F_3 = p_1(A_1 - A_2) = p_1\pi d^2/4 \quad (2\text{-}18)$$

$$v_3 = 4q/\pi d^2 \quad (2\text{-}19)$$

由式(2-18)和式(2-19)可知,差动连接时液压缸的推力比非差动连接时小,速度比非差动连接时要大,利用这一点,可使在不加大油源流量的情况下,得到较快的运动速度,这种连接方式被广泛应用于组合机床的液压动力系统及其他机械设备的快速运动中。

2. 柱塞缸

图 2-2-4(a)所示为柱塞缸,其只能实现一个方向的液压传动,反向运动需要依靠外力。如若需要实现双向运动,则必须成对使用。如图 2-2-4(b)所示,在液压缸中的柱塞和缸筒不接触的情况下,运动时由缸盖上的导向套来导向,因此这种缸筒的内壁不需精加工,就特别适

用于行程较长的场合。

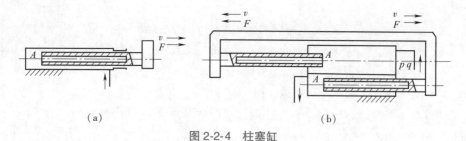

图 2-2-4 柱塞缸

柱塞缸输出的推力和速度各为

$$F = p_A = p\pi d^2/4 \qquad (2\text{-}20)$$
$$v_1 = q/A = 4q/\pi d^2 \qquad (2\text{-}21)$$

3. 其他液压缸

(1) 增压液压缸如图 2-2-5 所示。增压液压缸又称增压器，它利用活塞和柱塞有效面积的不同使液压系统中的局部区域获得高压。增压液压缸有单作用和双作用两种型式。单作用增压缸的工作原理如图 2-2-5(a)所示，当输入活塞缸的液体压力为 p_1，活塞直径为 D，柱塞直径为 d 时，柱塞缸中输出的液体压力为高压。其值为

$$p_2 = p_1(D/d)^2 = Kp_1 \qquad (2\text{-}22)$$

式中，$K = (D/d)^2$ 称为增压比，它代表其增压程度。

显然增压能力是在降低有效能量的基础上得到的，也就是说增压缸仅仅是增大输出的压力，并不能增大输出的能量。

单作用增压缸在柱塞运动到终点时，不能再输出高压液体，需要将活塞退回到左端位置，再向右行时才又输出高压液体。为了克服这一缺点，可采用双作用增压缸，如图 2-2-5(b)所示，由两个高压端连续向系统供油。

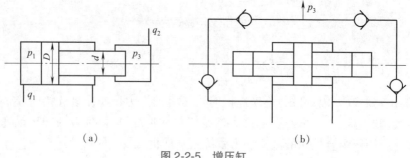

图 2-2-5 增压缸

(2) 伸缩缸。伸缩缸由两个或多个活塞缸套装而成，前一级活塞缸的活塞杆内孔是后一级活塞缸的缸筒，伸出时可获得很长的工作行程，缩回时可保持很小的结构尺寸，伸缩缸被广

泛用于起重运输车辆上。

伸缩缸可以是如图 2-2-6(a) 所示的单作用式,也可以是如图 2-2-6(b) 所示的双作用式,前者靠外力回程,后者靠液压回程。

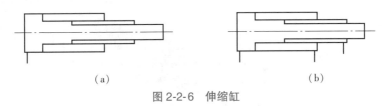

图 2-2-6 伸缩缸

伸缩缸的外伸动作是逐级进行的。首先是最大直径的缸筒以最低的油液压力开始外伸,当到达行程终点后,稍小直径的缸筒开始外伸,直径最小的末级最后伸出。随着工作级数变大,外伸缸筒直径越来越小,工作油液压力随之升高,工作速度变快。

(3)齿轮缸如图 2-2-7 所示。它由两个柱塞缸和一套齿条传动装置组成。柱塞的移动经齿轮齿条传动装置变成齿轮的传动,用于实现工作部件的往复摆动或间歇进给运动。

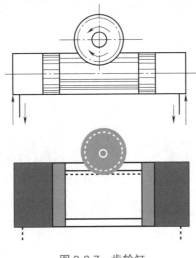

图 2-2-7 齿轮缸

三、液压缸的典型结构

图 2-2-8 所示为一个比较常见的双作用单活塞杆液压缸。它是由缸底 20、缸盖兼导向套 9、缸筒 10、活塞 11 和活塞杆 18 组成。缸筒一端与缸底焊接,另一端缸盖(导向套)与缸筒用卡键 6、套 5 和弹簧挡圈 4 固定,以便于拆装检修,两端设有油口 A 和 B。活塞 11 与活塞杆 18 利用卡键 15、卡键帽 16 和弹簧挡圈 17 连接在一起。活塞与缸孔的密封采用的是一对 Y 形聚氨酯密封圈 12,由于活塞与缸孔有一定间隙,采用由尼龙 1010 制成的耐磨环(又称支承环)13 定心导向,活塞杆 18 和活塞 11 的内孔由 O 形密封圈 14 密封。较长的导向套 9 则可保证活

塞杆不偏离中心，导向套外径由 O 形密封圈 7 密封，而其内孔则由 Y 形密封圈 8 可防止油外漏，防尘圈 3 可有效避免将灰尘带入缸内。缸与杆端销孔与外界连接，销孔内有尼龙衬套抗磨。

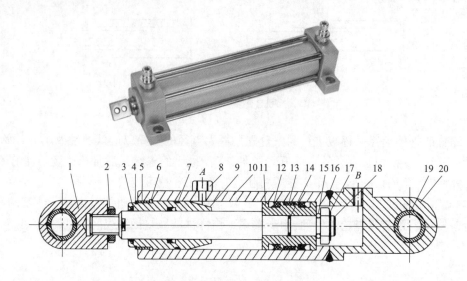

图 2-2-8　双作用单活塞杆液压缸

1—耳环；2—螺母；3—防尘圈；4、17—弹簧挡圈；5—套；6、15—卡键；7、14—O 形密封圈；8、12—Y 形密封圈；9—缸盖兼导向套；10—缸筒；11—活塞；13—耐磨环；16—卡键帽；18—活塞杆；19—衬套；20—缸底

四、液压缸的设计

液压缸是液压传动的执行元件，它与主机的工作机构有直接的联系。对于不同的主机机构，液压缸所要求的用途和工作要求也不同。因此，在设计液压缸之前，必须对整个液压系统进行工况分析，编制负载图，选定系统的工作压力，然后根据实际使用要求选择结构类型，按运动要求、负载情况、最大行程等确定其主要工作尺寸，进行强度、稳定性和缓冲验算，最后进行结构设计。

1. 液压缸的设计内容和步骤

(1) 选择液压缸的类型以及其他各部分结构形式。

(2) 确定液压缸的工作参数和结构尺寸。

(3) 结构强度、刚度的计算和校核。

(4) 导向、密封、防尘、排气和缓冲等装置的设计。

(5) 绘制装配图、零件图，编写设计说明书。

2. 计算液压缸的结构尺寸

液压缸的结构尺寸主要有缸筒内径 D、活塞杆外径 d 和缸筒长度 L。

(1) 缸筒内径 D。液压缸的缸筒内径是指根据负载的大小选定工作压力与往返运动速度比,以此计算出液压缸的有效工作面积,从而得到缸筒内径 D。然后,再从 GB/T 2348—2018 标准中选取最接近的标准值作为所设计的缸筒内径。

根据负载和工作压力的大小确定 D。

①无杆腔作工作腔时

$$D = \sqrt{\frac{4F_{max}}{\pi p_i}} \qquad (2\text{-}23)$$

②有杆腔作工作腔时

$$D = \sqrt{\frac{4F_{max}}{\pi p_i} + d^2} \qquad (2\text{-}24)$$

式中,p_i 为缸工作腔的工作压力,可根据机床类型或负载的大小来确定;F_{max} 为最大作用负载。

(2)活塞杆外径 d。活塞杆外径 d 通常先通过满足速度或速度比的要求后,再校核其结构的强度和稳定性。若速度比为 λ_v,则该处应有一个带根号的式子:

$$D = \sqrt{\frac{\lambda_v - 1}{\lambda_v}} \qquad (2\text{-}25)$$

也可根据活塞杆受力状况来确定,一般为受拉力作用时,$d = (0.3 \sim 0.5)D$。

受压力作用时:

$p_i < 5$ MPa 时,$d = (0.5 \sim 0.55)D$;

5 MPa $< p_i <$ 7 MPa 时,$d = (0.6 \sim 0.7)D$;

$p_i > 7$ MPa 时,$d = 0.7D$。

(3)缸筒长度 L。缸筒长度 L 由最大工作行程长度加上各种结构需要来确定,即

$$L = l + B + A + M + C$$

式中,l 为活塞的最大工作行程;B 为活塞宽度,一般为 $(0.6 \sim 1)D$;A 为活塞杆导向长度,取 $(0.6 \sim 1.5)D$;M 为活塞杆密封长度,由密封方式定;C 为其他长度。

一般缸筒的长度最好不超过内径的 20 倍。

另外,液压缸的结构尺寸设计中还要考虑最小导向长度 H。

3. 液压缸设计中应注意的问题

液压缸的设计和使用正确与否,直接影响其性能的发挥和发生故障可能性的大小。这方面常见的问题主要是液压缸设置不当、液压缸或活塞下垂、活塞杆承受偏载以及活塞杆的压杆失稳等。所以,在设计液压缸时,须注意以下几方面:

(1)尽量使液压缸的活塞杆在受拉状态下承受最大负载,或在受压状态下具有良好的稳定性。

(2)考虑液压缸行程终了处的制动问题和液压缸的排气问题。缸内如无缓冲装置和排气装置,系统中需有相应的措施。但并非所有的液压缸都需要考虑此问题。

(3)正确安装及固定液压缸。如承受弯曲的活塞杆不能用螺纹连接,要用止口连接;液压缸只能在一端定位,而不能在两端用键或销定位,目的是不致阻碍它在受热时的膨胀;如冲击

载荷使活塞杆压缩,定位件须设置在活塞杆端,反之则设置在缸盖端。

(4)液压缸各部分的结构需根据推荐的结构形式和设计标准进行设计,尽可能做到结构简单、紧凑,加工、装配和维修方便。

(5)在能够满足运动行程和负载力的条件下,应尽可能缩小液压缸的轮廓尺寸。

(6)要保证液压缸密封可靠、防尘良好。可靠的密封性是其正常工作的重要指标。如泄漏严重,不仅会降低液压缸的工作效率,甚至会导致其不能正常运行(如满足不了负载力和运动速度要求等)。良好的防尘措施,还有助于提高液压缸的使用寿命。

总之,液压缸的设计内容不是一成不变的,根据实际情况可稍做修改,也可增加一些新的内容。设计步骤也需要经过多次反复尝试修改,才能得到正确、合理的设计结果。在设计液压缸时,正确选择液压缸的类型是所有设计计算的前提基础。在选择液压缸的类型时,要从机器设备的行程长短、动作特点、运动性能等要求出发,同时还需要考虑主机的结构特征给液压缸提供的安装空间和具体位置。

例如,机器的往复直线运动直接采用液压缸来实现是最简单又方便的。对于要求往返运动速度一致的场合,可采用双活塞杆式液压缸。若有快速返回的需求,则宜用单活塞杆式液压缸,并可考虑用差动连接。行程较长时,可采用柱塞缸,以减少加工的困难。行程较长但负载不大时,也可考虑采用一些传动装置来扩大行程。往复摆动运动既可用摆动式液压缸,也可用直线式液压缸加连杆机构或齿轮-齿条机构来实现。

任务实施

1. 画出设备上的液压缸的图形符号。
2. 能熟练进行液压缸的缸径设计。
3. 分组讨论,总结出各种液压缸的不同应用场景。

思考与练习

1. 液压执行元件有哪些?它们的基本原理是什么?
2. 液压马达和液压泵有什么联系和不同?
3. 简述液压缸的特点及应用。

项目 3
液压控制阀

任务 1　方向控制阀的认知

任务描述

了解常见方向控制阀的基本工作原理及符号,熟悉方向控制阀的种类及常见方向控制阀的不同应用场景。

相关知识

液压阀的用途是控制液压系统中油液的流动方向或调节其压力和流量,它可分为方向阀、压力阀和流量阀三大类。一个形状相同的阀,可以因为作用机制的不同,而具有不同的功能。压力阀和流量阀利用通流截面的节流作用控制系统的压力和流量,而方向阀则利用通油通道的更换控制油液的流动方向。这就是说,尽管液压阀存在着各种各样不同的类型,它们之间保持着一些基本共同之处。例如:

在结构上,所有的阀都由阀体、阀芯(转阀或滑阀)和驱使阀芯动作的元、部件(如弹簧、电磁铁)组成。

在工作原理上,所有阀的开口大小,阀进、出口间压差以及流过阀的流量之间的关系都符合孔口流量公式,仅是各种阀的控制参数各不相同而已。

方向控制阀是用来改变液压系统中各油路之间液流通断关系的阀类,包括单向阀、换向阀及压力表开关等。

一、单向阀

液压系统中常见的单向阀有普通单向阀和液控单向阀两种。

1. 普通单向阀

普通单向阀的作用,是使油液只能沿一个方向流动,不允许反向倒流。图 3-1-1(a)所示为一种管式普通单向阀的结构。压力油从阀体左端的通口 P_1 流入时,克服弹簧 3 作用在阀芯 2 上的力,使阀芯向右移动,打开阀口,并通过阀芯 2 上的径向孔 a、轴向孔 b 从阀体右端的通口流出。但是压力油从阀体右端的通口 P_2 流入时,它和弹簧力一起使阀芯锥面压紧在阀座上,使阀口关闭,油液无法通过。图 3-1-1(b)所示为单向阀的职能符号,图 3-1-1(c)为其实物图。

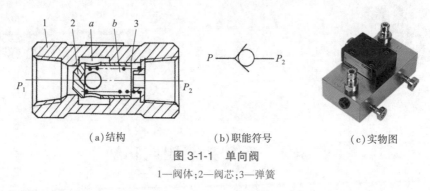

(a)结构　　　　(b)职能符号　　　　(c)实物图

图 3-1-1　单向阀

1—阀体;2—阀芯;3—弹簧

2. 液控单向阀

图 3-1-2(a)所示是液控单向阀的结构。当控制口 K 处无压力油通入时,它的工作机制和普通单向阀一样;压力油只能从通口 P_1 流向通口 P_2,不能反向倒流。当控制口 K 有控制压力油时,因控制活塞 1 右侧 a 腔通泄油口,活塞 1 右移,推动顶杆 2 顶开阀芯 3,使通口 P_1 和 P_2 接通,油液就可在两个方向自由通流。图 3-1-2(b)所示为液控单向阀的职能符号,图 3-1-2(c)为其实物图。

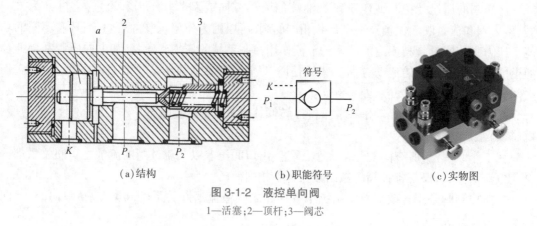

(a)结构　　　　(b)职能符号　　　　(c)实物图

图 3-1-2　液控单向阀

1—活塞;2—顶杆;3—阀芯

二、换向阀

换向阀利用阀芯相对于阀体的相对运动,使油路接通、关断,或变换油流的方向,从而使液压执行元件启动、停止或变换运动方向。换向阀的分类见表 3-1-1。

表 3-1-1 换向阀的分类

分类方式	类型
按阀芯的运动形式	滑阀、转阀
按阀的工作位置和通路数	二位二通、二位三通、二位四通、三位四通、三位五通等
按阀的操纵方式	手动、机动、电动、液动、电液动

1. 换向阀的工作原理

图 3-1-3 所示为换向阀的工作原理。在图示状态下,液压缸两腔不通压力油,活塞处于停止状态。若使阀芯 1 左移,阀体 2 的油口 P 和 A 连通,B 和 T 连通,则压力油经 P、A 进入液压缸左腔,右腔油液经 B、T 流回油箱,活塞向右运动;反之,若使阀芯右移,则油口 P 和 B 连通、A 和 T 连通,活塞便向左运动。

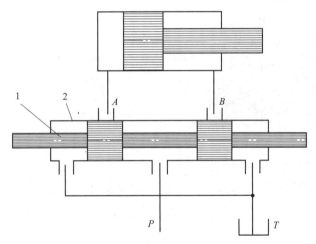

图 3-1-3 换向阀的工作原理
1—阀芯;2—阀体

换向阀的结构原理、图形符号及使用场合简述见表 3-1-2。

表 3-1-2 换向阀的结构原理、图形符号及使用场合

名称	结构原理图	图形符号	使用场合
二位二通换向阀	(图)	(图)	控制油液的连接与通断(相当于一个开关)
二位三通换向阀	(图)	(图)	控制油液方向(从一个方向变成另一个方向)

续上表

名　称	结构原理图	图形符号	使用场合
二位四通换向阀	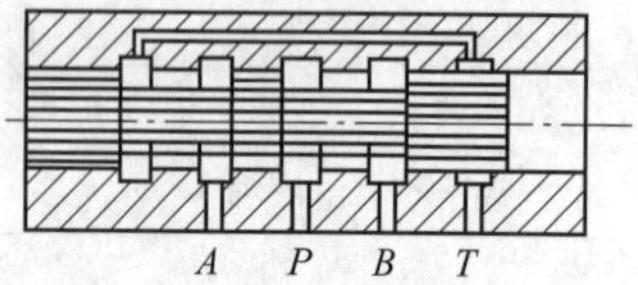 结构图（油口 A P B T）	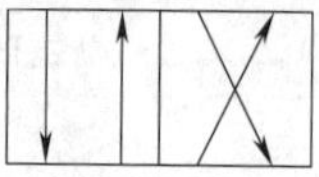 符号图	不能使执行元件在任意位置停止运动
二位五通换向阀	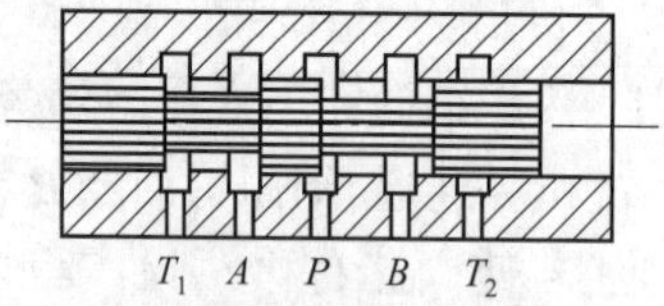 结构图（油口 T_1 A P B T_2）	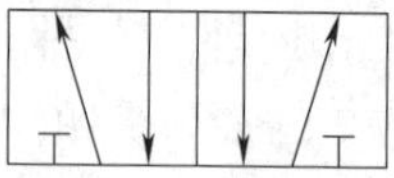 符号图	不能使执行元件在任意位置停止运动
三位四通换向阀	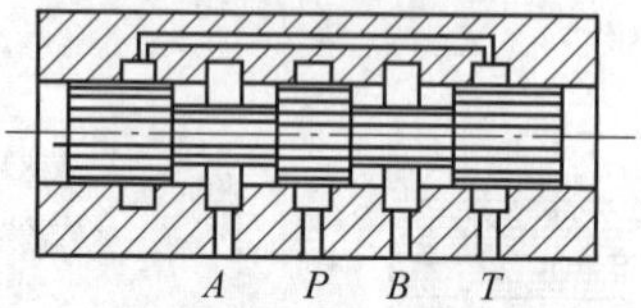 结构图（油口 A P B T）	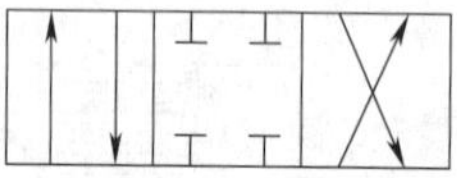 符号图	能使执行元件在任意位置停止运动
三位五通换向阀	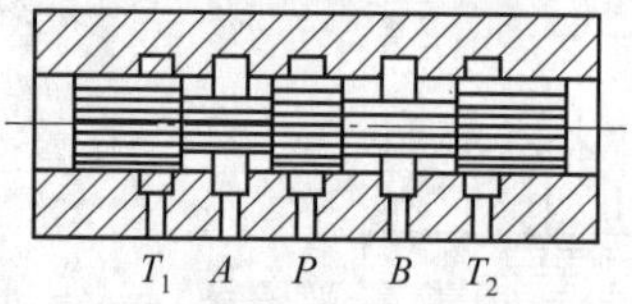 结构图（油口 T_1 A P B T_2）	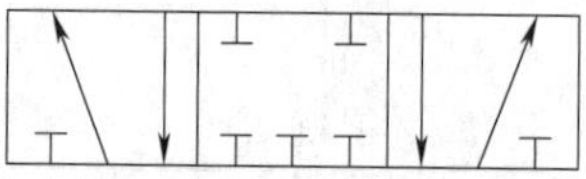 符号图	能使执行元件在任意位置停止运动

2. 换向阀的中位机能

换向阀处于常态位置时，其各油口的连通方式称为滑阀机能。三位换向阀的常态为中位，因此三位换向阀的滑阀机能又称中位机能。不同中位机能的三位换向阀阀体通用，仅阀芯台肩的结构、尺寸及内部通孔情况有区别，但作用区别又极大。三位四通换向阀常用的五种中位机能见表3-1-3。

表3-1-3　三位四通换向阀常用的五种中位机能

代号	结构简图	中位符号	中位油口的状态和特点
O	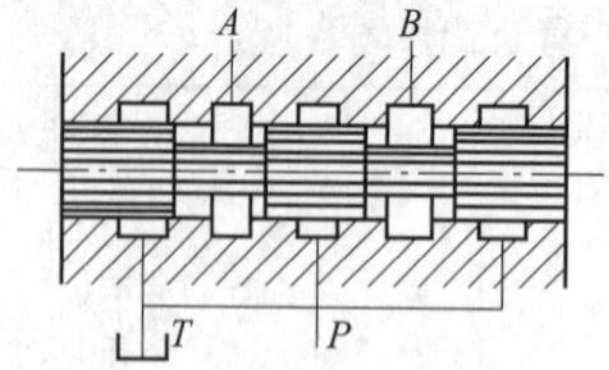 结构图（A B T P）	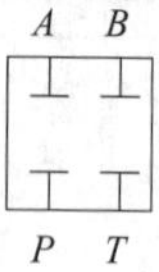 符号图（A B P T）	各油口全封闭，换向精度高，但有冲击，缸被锁紧，泵不卸荷，并联的缸可运动
H	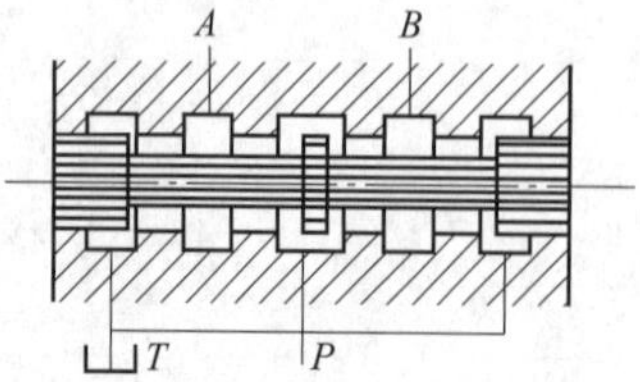 结构图（A B T P）	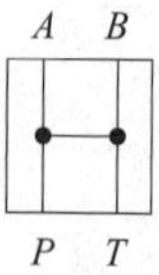 符号图（A B P T）	各油口全通，换向平稳，缸浮动，泵卸荷，其他缸不能并联使用

续上表

代号	结构简图	中位符号	中位油口的状态和特点
Y			P口封闭,A、B、T口相通,换向较平稳,缸浮动,泵不卸荷,并联的缸可运动
P			T口封闭,P、A、B口相通,换向最平稳,双杆缸浮动,单杆缸差动,泵不卸荷,并联的缸可运动
M			P、T口相通,A、B口封闭,换向精度高,但有冲击,缸被锁紧,泵卸荷,其他缸不能并联使用

在分析和选择阀的中位机能时,通常考虑以下几点：

(1) 系统保压。当P口被堵塞,系统保压,液压泵能用于多缸系统。当P口不太通畅地与T口接通时,系统能保持一定的压力,供控制油路使用。

(2) 系统卸荷。P口通畅地与T口接通时,系统卸荷。

(3) 启动平稳性。阀芯处于中位时,液压缸某腔如通油箱,则启动时该腔内因无油液起缓冲作用,启动不太平稳。

(4) 液压缸"浮动"和在任意位置上的停止。阀芯处于中位,当A、B两口互通时,卧式液压缸呈"浮动"状态,可利用其他机构移动工作台,调整其位置。当A、B两口堵塞或与P口连接(在非差动情况下),则可使液压缸在任意位置处停下来。三位五通换向阀的机能与上述相仿。

3. 换向阀的结构

在液压传动系统中广泛采用的是滑阀式换向阀,在这里主要介绍这种换向阀的几种典型结构。

(1) 手动换向阀。图3-1-4(a)为自动复位式手动换向阀。放开手柄1时,阀芯2在弹簧3的作用下自动回复中位。该阀适用于动作频繁、工作持续时间短的场合,操作比较完全,常用于工程机械的液压传动系统中。

如果将该阀的阀芯右端弹簧3的部位改为可自动定位的结构形式,即成为可在三个位置定位的手动换向阀。图3-1-4(b)为其职能符号。

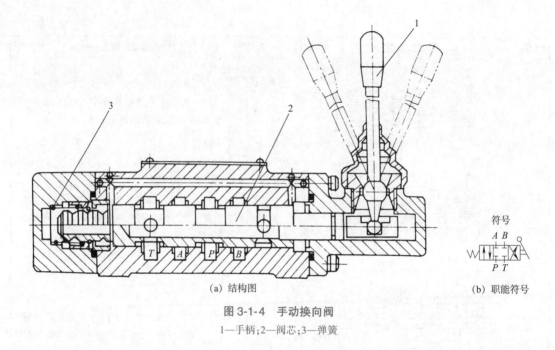

(a) 结构图 (b) 职能符号

图 3-1-4　手动换向阀
1—手柄；2—阀芯；3—弹簧

(2) 机动换向阀。机动换向阀又称行程阀，主要用来控制机械运动部件的行程。它是借助于安装在工作台上的挡铁 5 迫使阀芯移动，从而控制油液的流动方向。机动换向阀通常是二位的，有二通、三通、四通和五通几种，其中二位二通机动阀又分常闭式和常开式两种。图 3-1-5(a) 为滚轮式二位三通常闭式机动换向阀。在图示位置阀芯 2 被弹簧 1 压向上端，油腔 P 和 A 通，B 口关闭。当挡铁 5 压住滚轮 4，使阀芯 2 移动到下端时，就使油腔 P 和 A 断开，P 和 B 接通，A 口关闭。图 3-1-5(b) 所示为其职能符号。

(3) 电磁换向阀。电磁换向阀利用电磁铁的通电吸合与断电释放而直接推动阀芯来控制液流方向，是在电气系统与液压系统之间信号转换功能的元件。

电磁铁按使用电源的不同，可分为交流和直流两种。按衔铁工作腔是否有油液又可分为"干式"和"湿式"。交流电磁铁启动力较大，不需要专门的电源，吸合、释放快，动作时间为 0.01~0.03 s，其缺点是若电源电压下降 15% 以上，则电磁铁吸力明显减小。若衔铁不动作，干式电磁铁会在 10~15 min 后烧坏线圈(湿式电磁铁为 1~1.5 h)，且冲击及噪声较大，寿命低，因而在实际使用中交流电磁铁允许的切换频率一般为 10 次/min，不得超过 30 次/min。直流电磁铁工作较可靠，吸合、释放动作时间为 0.05~0.08 s，允许使用的切换频率较高，一般可达 120 次/min，最高可达 300 次/min，且冲击小、体积小、寿命长，但需有专门的直流电源，成本较高。此外，还有一种整体电磁铁，其电磁铁是直流的，但电磁铁本身带有整流器，通入的交流电经整流后再供给直流电磁铁。目前，新发展了一种油浸式电磁铁，不但衔铁，而且激磁线圈也都浸在油液中工作。它具有寿命更长，工作更平稳可靠等特点，但由于造价较高，应用面不广。

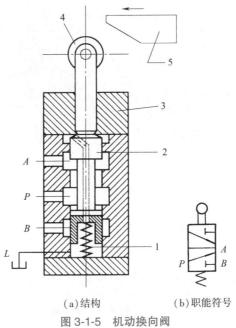

(a)结构　　　　(b)职能符号

图 3-1-5　机动换向阀

1—弹簧；2—阀芯；3—阀体；4—滚轮；5—挡铁

图 3-1-6(a)所示为二位三通交流电磁换向阀结构,在图示位置,油口 P 和 A 相通,油口 B 断开;当电磁铁通电吸合时,推杆 1 将阀芯 2 推向右端,这时油口 P 和 A 断开,而与 B 相通。而当磁铁断电释放时,弹簧 3 推动阀芯复位。图 3-1-6(b)所示为其职能符号。

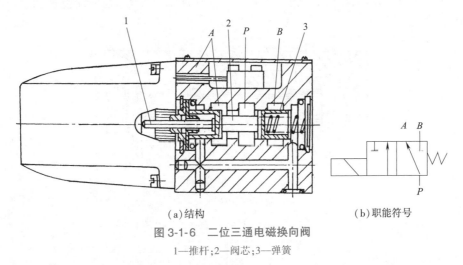

(a)结构　　　　(b)职能符号

图 3-1-6　二位三通电磁换向阀

1—推杆；2—阀芯；3—弹簧

如前所述,电磁换向阀就其工作位置来说,有二位和三位等。二位电磁阀有一个电磁铁靠弹簧复位;三位电磁阀有两个电磁铁。图 3-1-7 所示为一种三位五通电磁换向阀的结构和职能符号。

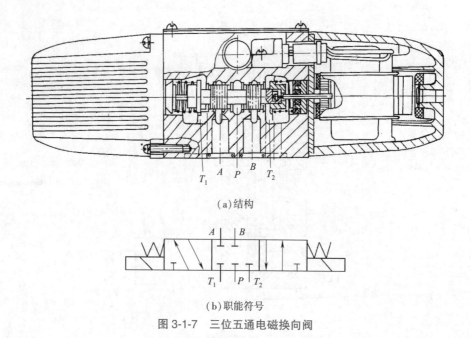

(a)结构

(b)职能符号

图 3-1-7　三位五通电磁换向阀

(4)液动换向阀。液动换向阀是利用控制油路的压力油来改变阀芯位置的换向阀，图 3-1-8 所示为三位四通液动换向阀的结构和职能符号。阀芯是由其两端密封腔中油液的压差来移动的，当控制油路的压力油从阀右边的控制油口 K_2 进入滑阀右腔时，K_1 接通回油，阀芯向左移动，使压力油口 P 与 B 相通，A 与 T 相通；当 K_1 接通压力油，K_2 接通回油时，阀芯向右移动，使得 P 与 A 相通，B 与 T 相通；当 K_1、K_2 都通回油时，阀芯在两端弹簧和定位套作用下回到中间位置。

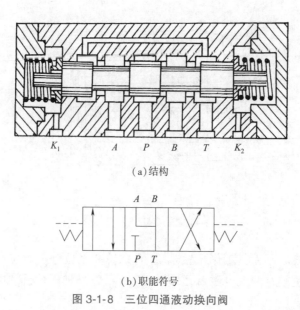

(a)结构

(b)职能符号

图 3-1-8　三位四通液动换向阀

(5)电液换向阀。在大中型液压设备中,当通过阀的流量较大时,作用在滑阀上的摩擦力和液动力较大,此时电磁换向阀的电磁铁推力相对太小,需要用电液换向阀来代替电磁换向阀。电液换向阀是由电磁滑阀和液动滑阀组合而成。电磁滑阀起先导作用,它可以改变控制液流的方向,从而改变液动滑阀阀芯的位置。由于操纵液动滑阀的液压推力可以很大,所以主阀芯的尺寸可以做得很大,允许有较大的油液流量通过。这样用较小的电磁铁就能控制较大的液流。

图 3-1-9 所示为弹簧对中型三位四通电液换向阀的结构和职能符号。当先导电磁阀左边的电磁铁通电后使其阀芯向右边位置移动,来自主阀 P 口或外接油口的控制压力油可经先导电磁阀的 A' 口和左单向阀进入主阀左端容腔,并推动主阀阀芯向右移动。这时,主阀阀芯右端容腔中的控制油液可通过右边的节流阀经先导电磁阀的 B' 口和 T' 口,再从主阀的 T 口或外接油口流回油箱(主阀阀芯的移动速度可由右边的节流阀调节),使主阀 P 与 A、B 和 T 的油路相通;反之,由先导电磁阀右边的电磁铁通电,可使 P 与 B、A 与 T 的油路相通;当先导电磁阀的两个电磁铁均不带电时,先导电磁阀阀芯在其对中弹簧作用下回到中位。此时,来自主阀 P 口或外接油口的控制压力油不再进入主阀芯的左、右两容腔,主阀芯左右两腔的油液通过先导电磁阀中间位置的 A'、B' 两个油口与先导电磁阀 T' 口相通,如图 3-1-9(b)所示,再从主阀的 T 口或外接油口流回油箱。主阀阀芯在两端对中弹簧的预压力的推动下,依靠阀体定位,准确地回到中位,此时主阀的 P、A、B 和 T 油口均不通。电液换向阀除了上述的弹簧对中以外还有液压对中的,在液压对中的电液换向阀中,先导式电磁阀在中位时,A'、B' 两油口均与油口 P 连通,而 T' 则封闭,其他方面与弹簧对中的电液换向阀基本相似。

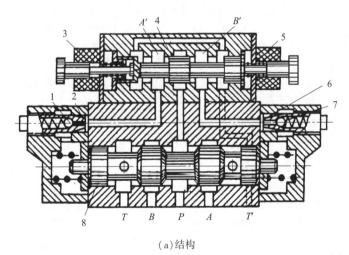

(a)结构

图 3-1-9 电液换向阀

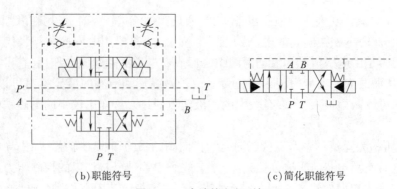

(b)职能符号　　　　　　　　　　　(c)简化职能符号

图 3-1-9　电液换向阀(续)

1、6—节流阀；2、7—单向阀；3、5—电磁铁；4—电磁阀阀芯；8—主阀阀芯

4. 方向控制阀的常见故障及其排除方法

方向阀常见故障及排除方法见表 3-1-4。

表 3-1-4　方向阀常见故障及排除方法

故障现象	原因分析	排除方法
阀芯不动或不到位	(1)滑阀卡住 ①滑阀(阀芯)与阀体配合间隙过小,阀芯在孔中容易卡住而不能动作或动作不灵 ②阀芯(或阀体)被碰伤,油液被污染 ③阀芯几何形状超差。阀芯与阀孔装配不同心,产生轴向液压卡紧现象 (2)液动换向阀的控制油路有故障 ①油液控制压力不够,滑阀不动,不能换向或换向不到位 ②节流阀关闭或堵塞 ③滑阀两端泄油口没有接回油箱或泄油管堵塞 (3)电磁铁故障 ①交流电磁铁因滑阀卡住,铁芯吸不到底面而被烧毁 ②漏磁、吸力不足 ③电磁铁接线焊接不良,接触不好 (4)弹簧折断、漏装、太软,都不能使滑阀恢复中位,因而不能换向电磁换向阀的推杆磨损后长度不够或行程不正确,使阀芯移动过小或过大,都会引起换向不灵或不到位	(1)检修滑阀 ①检查间隙情况,研修或更换阀芯 ②检查、修磨或重配阀芯,必要时更换新油 ③检查、修正几何偏差及同心度,检查液压卡紧情况并修复 (2)检查控制油路 ①提高控制油压,检查弹簧是否过硬,以便更换 ②检查、清洗节流口 ③检查并接通回油箱,清洗回油管,使之畅通 (3)检查并修复 ①检查滑阀卡住故障,更换电磁铁 ②检查漏磁原因,更换电磁铁 ③检查并重新焊接 (4)检查、更换或补装 检查并修复,必要时可更换推杆
换向冲击与噪声	(1)控制流量过大,滑阀移动速度太快,产生冲击声 (2)单向节流阀芯与阀孔配合间隙过大,单向阀弹簧漏装,阻尼失效,产生冲击声 (3)电磁铁的铁芯接触面不平或接触不良 (4)滑阀时卡时动或局部摩擦力过大 (5)固定电磁铁的螺栓松动而产生振动	(1)调小单向节流阀的节流口,减慢滑阀移动速度 (2)检查、修整(修复)到合理间隙,补装弹簧 (3)清除异物,并修整电磁铁的铁芯 (4)研磨修整或更换滑阀 (5)紧固螺栓,并加防松垫圈

任务实施

1. 画出设备上的方向控制阀的图形符号。
2. 能熟练调节方向控制阀进行油路的换向。
3. 分组讨论,总结各种方向控制阀的应用场景。

任务2 压力控制阀的认知

任务描述

了解常见压力控制阀的基本工作原理及符号,熟悉压力控制阀的种类及常见压力控制阀的不同应用场景。

相关知识

在液压传动系统中,控制油液压力高低的液压阀称为压力控制阀,简称压力阀。这类阀的共同点是利用作用在阀芯上的液压力和弹簧力相平衡的原理工作。

在各种液压系统中,工作需要不同,对压力控制的要求也不相同:有的需要限制液压系统的最高压力,如安全阀;有的需要稳定液压系统中某处的压力值(或者压力差、压力比等),如溢流阀、减压阀等定压阀;还有的是利用液压力作为信号控制其动作,如顺序阀、压力继电器等。

一、溢流阀

1. 溢流阀的结构和工作原理

常用的溢流阀按其结构形式和基本动作方式可归结为直动式和先导式两种。
(1)直动式溢流阀。

直动式溢流阀依靠系统中的压力油直接作用在阀芯上与弹簧力相平衡,控制阀芯的启闭动作。图3-2-1(a)所示为一种低压直动式溢流阀,P 是进油口,T 是回油口。进口压力油经阀芯3中间的阻尼孔 a 作用在阀芯的底部端面上。当进油压力较小时,阀芯在弹簧2的作用下处于下端位置,将 P 和 T 两油口隔开。当油压力升高,在阀芯下端所产生的作用力超过弹簧的压紧力 F。此时,阀芯上升,阀口被打开,将多余的油液排回油箱。阀芯上的阻尼孔 a 的作用是对阀芯的动作产生阻尼,以提高阀的工作平衡性。调整螺母1可以改变弹簧的压紧力,这样也就调整了溢流阀进口处的油液压力 p。

这种低压直动式溢流阀一般用于压力小于 2.5 MPa 的小流量场合。图3-2-1(b)所示为直动式溢流阀的图形符号。由图3-2-1(a)还可看出,在常位状态下,溢流阀进出油口之间是不相通的,而且作用在阀芯上的液压力是由进口油液压力产生的,经溢流阀芯的泄漏油液经内泄漏通道进入回油口 T。

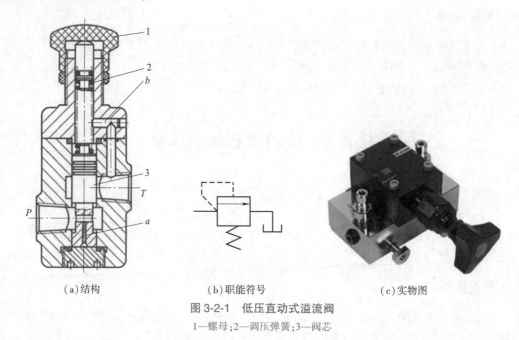

(a)结构　　　　(b)职能符号　　　　(c)实物图

图 3-2-1　低压直动式溢流阀
1—螺母；2—调压弹簧；3—阀芯

(2)先导式溢流阀。

图 3-2-2 所示为先导式溢流阀的结构示意图。在图中，压力油从 P 口进入，通过阻尼孔 e 后作用在锥阀芯 3 上。当进油口压力较低，导阀上的液压作用力不足以克服导阀右边的弹簧 2 的作用力时，导阀关闭，没有油液流过阻尼孔，所以锥阀芯 3 两端压力相等。在较软的主阀弹簧 2 作用下阀芯 3 处于最下端位置，溢流阀阀口 P 和 T 隔断，没有溢流。当进油口压力升高到作用在导阀上的液压力大于导阀弹簧作用力时，导阀打开，压力油就可通过阻尼孔、经导阀流回油箱。由于阻尼孔的作用，使主阀芯上端的液压力 p_2 小于下端压力 p_1，当这个压力差作用在主阀芯上的力等于或超过主阀预设的弹簧力 F_s，主阀芯开启，油液从 P 口流入，经主阀阀口由 T 流回油箱，实现溢流。

先导式溢流阀有一个远程控制口 K。如果将 K 口用油管接到另一个远程调压阀（远程调压阀的结构和溢流阀的先导控制部分一样），调节远程调压阀的弹簧力，即可调节溢流阀主阀芯上端的液压力，从而对溢流阀的溢流压力实现远程调压。但是，远程调压阀所能调节的最高压力不得超过溢流阀本身导阀的调整压力。当远程控制口 K 通过二位二通阀接通油箱时，主阀芯上端的压力接近于零。由于主阀弹簧较软，主阀芯上移到最高位置，阀口开得很大。这时溢流阀 P 口处压力很低，系统的油液在低压下通过溢流阀流回油箱，实现卸荷。

2. 溢流阀的作用

根据溢流阀在液压系统中所起的作用，溢流阀除可溢流外，还可作安全阀、卸荷阀和背压阀使用。

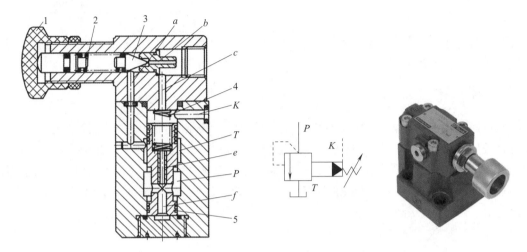

图 3-2-2 先导式溢流阀

1—调节螺母；2—调压弹簧；3—锥阀芯；4—弹簧；5—主阀芯

(1) 作溢流阀用。

在采用定量泵供油的液压系统中，由流量控制阀调节进入执行元件的流量，定量泵输出的多余油液则从溢流阀流回油箱。在工作过程中溢流阀口常开，系统的工作压力由溢流阀调整并保持基本恒定。图 3-2-3(a) 所示为溢流阀。

(2) 作安全阀用。

图 3-2-3(b) 所示为一变量泵供油系统，执行元件的速度由变量泵自身调节，系统中无多余油液，系统工作压力随负载变化而变化。正常工作时，溢流阀口关闭。一旦过载，溢流阀口立即打开，使油液流回油箱，系统压力不再升高，以保障系统安全。

(3) 作卸荷阀用。

如图 3-2-3(c) 所示，将先导式溢流阀远程控制口 K 通过二位二通电磁阀与油箱连接。当电磁铁断电时，远程控制口 K 被堵塞，溢流阀起溢流稳压作用。当电磁铁通电时，远程控制口 K 通油箱，溢流阀的主阀芯上端压力接近于零。此时，溢流阀口全开，回油阻力很小，泵输出的油液便在低压下经溢流阀口流回油箱，使液压泵卸荷，从而减小系统功率损失，故溢流阀起卸荷作用。

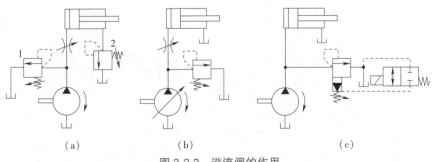

图 3-2-3 溢流阀的作用

（4）作背压阀用。

如图 3-2-3(a) 所示，溢流阀 2 接在回油路上，可对回油产生阻力，即形成背压。利用背压可提高执行元件的运动平稳性。

二、减压阀

减压阀是使出口压力（二次压力）低于进口压力（一次压力）的一种压力控制阀。其作用是降低液压系统中某一回路的油液压力，使用一个油源能同时提供两个或几个不同压力的输出。减压阀在各种液压设备的夹紧系统、润滑系统和控制系统中应用较多。此外，当油液压力不稳定时，在回路中串入一减压阀可得到一个稳定的较低的压力。减压阀也有直动式和先导式之分，先导式减压阀应用较多。

1. 减压阀的工作原理

图 3-2-4(b) 所示为先导式减压阀的外形。其结构原理如图 3-2-4(a) 所示，它在结构上和先导式溢流阀类似，也是由先导阀和主阀两部分组成。压力油从阀的进油口进入进油腔 P_1，经开口为 x 的减压阀减压后，再从出油腔 P_2 和出油口流出。出油腔压力油经小孔 f 进入主阀芯 5 的下端，同时经阻尼孔 e 流入主阀芯上端，再经孔 c 和 b 作用于锥阀芯 3 上。当出油口压力较低时，先导阀关闭，主阀芯两端压力相等，主阀芯被平衡弹簧 4 压在最下端（图示位置），减压阀口开度为最大，压降为最小，减压阀不起减压作用。当出油口压力达到先导阀的调定压力时，先导阀开启，此时 P_2 腔的部分压力油经孔 e、c、b、先导阀口、孔 a 和泄漏口 L 流回油箱。由于阻尼孔 e 的作用，主阀芯两端产生压力差，主阀芯便在此压力差作用下克服平衡弹簧的弹力上移，减压阀口减小，使出油口压力降低至调定压力。若由于外界干扰（如负载变化）而使出油口压力变化，则减压阀会自动调整减压阀口的开度以保持出油压力稳定。调节螺母 1 即可调节调压弹簧 2 的预压缩量，从而调定减压阀出油口压力。中压先导式减压阀的调压范围为 2.5~8.0 MPa，适用于中、低压系统。图 3-2-4(c) 所示为直动式减压阀的图形符号，也是减压阀的一般符号。图 3-2-4(d) 所示为先导式减压阀的图形符号，可仿前述先导式溢流阀来推演，这里不再赘述。

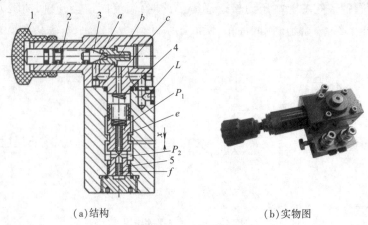

(a) 结构　　　　　　(b) 实物图

图 3-2-4　减压阀

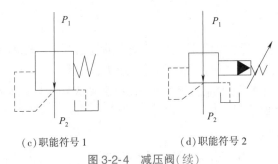

(c)职能符号1　　　　　(d)职能符号2

图 3-2-4　减压阀(续)

1—调节螺母；2—阻尼孔 x；3—锥阀芯；4—平衡弹簧；5—主阀芯

将先导式减压阀和先导式溢流阀进行比较,它们之间有如下几点不同之处：

①减压阀保持出口压力基本不变,而溢流阀保持进口处压力基本不变。

②在不工作时,减压阀进、出油口互通,而溢流阀进出油口不通。

③为保证减压阀出口压力调定值恒定,它的导阀弹簧腔需通过泄油口单独外接油箱；而溢流阀的出油口是通油箱的,所以它的导阀的弹簧腔和泄漏油可通过阀体上的通道和出油口相通,不必单独外接油箱。

2. 减压阀的应用

减压阀在夹紧油路、控制油路和润滑油路中应用较多。图 3-2-5 所示为减压阀用于夹紧油路的原理图,液压泵除供给主油路压力油外,还经分支油路上的减压阀为夹紧缸提供比液压泵供油压力低的稳定压力油,其夹紧力大小由减压阀来调节控制。

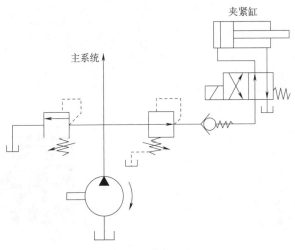

图 3-2-5　减压阀的应用

三、顺序阀

顺序阀是用来控制液压系统中各执行元件动作的先后顺序。依控制压力的不同,顺序阀又可分为内控式和外控式两种。前者用阀的进口压力控制阀芯的启闭,后者用外来的控制压力油控制阀芯的启闭(即液控顺序阀)。顺序阀也有直动式和先导式两种,前者一般用于低压系统,后者用于中高压系统。

1. 顺序阀的工作原理

图 3-2-6 所示为直动式顺序阀的工作原理图和图形符号。当进油口压力 p_1 较低时,阀芯在弹簧作用下处下端位置,进油口和出油口不相通。当作用在阀芯下端的油液的液压力大于弹簧的预紧力时,阀芯向上移动,阀口打开,油液便经阀口从出油口流出,从而操纵另一执行元件或其他元件动作。由图可见,顺序阀和溢流阀的结构基本相似,不同的只是顺序阀的出油口通向系统的另一压力油路,而溢流阀的出油口通油箱。此外,由于顺序阀的进、出油口均为压力油,所以它的泄油口 L 必须单独外接油箱。因顺序阀的控制油液直接从进油口引入,故称为内控外泄式顺序阀。

将图 3-2-6(a)中的下盖 1 旋转 90°或 180°安装,切断原控油路,将外控口 K 的螺塞取下,接通控制油路,则阀的开启由外部压力油控制,便构成外控外泄式顺序阀,其图形符号如图 3-2-6(d)所示。若再将上盖 6 旋转 180°安装,并将外泄口 L 堵塞,则弹簧腔与出油口相通,构成外控内泄式顺序阀,其图形符号如图 3-2-6(e)所示。

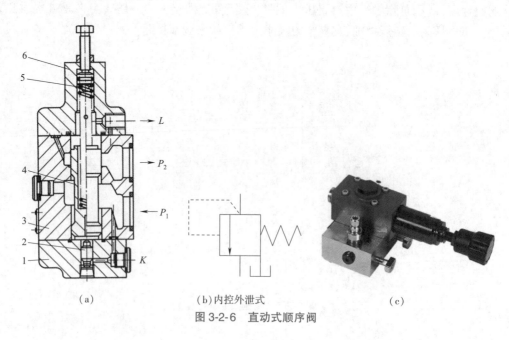

图 3-2-6 直动式顺序阀

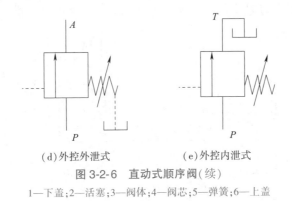

(d) 外控外泄式　　　(e) 外控内泄式

图 3-2-6　直动式顺序阀(续)

1—下盖；2—活塞；3—阀体；4—阀芯；5—弹簧；6—上盖

2. 顺序阀的应用

图 3-2-7 所示为机床夹具上用顺序阀实现定工件先定位，后夹紧的顺序动作回路。当换向阀 U 右位工作时，压力油，首先进入定位缸下腔，完成定位动作以后，系统压力升高。当达到顺序阀调定压力(为保证工作可靠，顺序阀的调定压力，应比定位缸最高工作压力高 0.5~0.8 MPa)时，顺序阀打开，压力油经顺序阀进入夹紧缸下腔，实现液压夹紧。当换向阀左位工作时，压力油同时进入定位缸和夹紧缸上腔，拔出定位销，松开工件，夹紧缸通过单向阀回油。此外，顺序阀还可用作卸荷阀、平衡阀和背压阀。

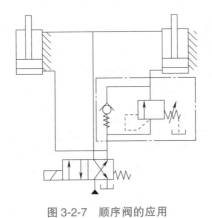

图 3-2-7　顺序阀的应用

四、压力继电器

压力继电器是一种将油液的压力信号转换成电信号的电液控制元件，当油液压力达到压力继电器的调定压力时，即发出电信号，以控制电磁铁、电磁离合器、继电器等元件动作，使油路卸压、换向、执行元件实现顺序动作，或关闭电动机，使系统停止工作，起安全保护作用等。

图 3-2-8 所示为常用柱塞式压力继电器的结构示意图和职能符号。如图所示，当从压力继电器下端进油口通入的油液压力达到调定压力值时，推动柱塞 1 上移，此位移通过杠杆 2 放大

后推动开关4动作。改变弹簧3的压缩量即可以调节压力继电器的动作压力。

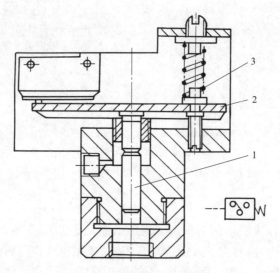

图 3-2-8　柱塞式压力继电器
1—柱塞；2—杠杆；3—弹簧；4—开关

五、压力控制阀的性能比较及常见故障排除方法

各种压力控制阀的结构和原理相似，在结构上仅有局部不同，有的是进出油口连接差异，有的是阀芯结构形状有局部改变。熟悉各类压力控制阀的结构性能特点，对分析与排除故障很有帮助。表 3-2-1 列出了溢流阀、减压阀和顺序阀的性能比较，表 3-2-2 列出了先导式溢流阀的常见故障及其排除方法。

表 3-2-1　溢流阀、减压阀和顺序阀的性能比较

名　称	溢流阀	减压阀	顺序阀
职能符号			
控制油路特点	把进口油液引到阀芯底部，与弹簧力平衡，所以是控制进油路压力，常态下阀口关闭	把出油口油液引到阀芯底部，与弹簧力平衡，所以是控制出口油路压力，常态下阀口全开	同溢流阀，把进油口油液引到阀芯底部，所以是控制进油路压力，常态下阀口关闭
回油特点	阀的出油口油液直接流回油箱，故泄漏油可在阀体内与回油口连通，属于内泄式	阀的出油口油液是低于进油压力的二次压力油，供给辅助油路，所以要单独设置泄油口，属于外泄式	同减压阀，即出油口油液接另一个缸中，所以要单独设置泄油口，也属于外泄式

续上表

名称	溢流阀	减压阀	顺序阀
基本用法	用做溢流阀、安全阀、卸荷阀,一般接在泵的出口,与主油路并联。若作背压阀用,则串联在回油路上,调定压力较低	串联在系统内,接在液压泵与分支油路之间	串联在系统中,控制执行机构的顺序动作。多数与单向阀并联,作为单向顺序阀用
举例及说明	用做溢流阀时,油路常开,泵的压力取决于溢流阀的调整压力,多用于节流调速的定量泵系统用做安全阀时,油路常闭,系统压力超过安全阀的调定值时,安全阀打开,多用于变量泵系统	作减压用,使辅助油路获得比主油路低且较稳定的压力油,阀口是常开的	用作顺序阀、平衡阀。顺序阀结构与溢流阀相似,经过适当改装,两阀可以互相代替。但顺序阀要求密封性较高,否则会产生误动作

表 3-2-2 先导式溢流阀的常见故障及其排除方法

故障现象	原因分析	排除方法
无压力	(1)主阀芯阻尼孔堵塞 (2)主阀芯在开启位置卡死 (3)主阀平衡弹簧折断或弯曲而使主阀芯不能复位 (4)调压弹簧弯曲或漏装 (5)锥阀(或钢球)漏装或破碎 (6)先导阀阀座破碎 (7)远程控制口通油箱	(1)清洗阻尼孔,过滤或换油 (2)检修,重新装配(阀盖螺钉紧固力要均匀),过滤或换油 (3)换弹簧 (4)更换或补装弹簧 (5)补装或更换 (6)更换阀座 (7)检查电磁换向阀的工作状态或远程控制口的通断状态
压力波动大	(1)主阀芯动作不灵活,时有卡住现象 (2)主阀芯和先导阀阀座阻尼孔时堵时通 (3)弹簧弯曲或弹簧刚度太小 (4)阻尼孔太大,消振效果差 (5)调压螺母未锁紧	(1)修换阀芯,重新装配(阀盖螺钉紧固力应均匀),过滤或换油 (2)清洗缩小的阻尼孔,过滤或换油 (3)更换弹簧 (4)适当缩小阻尼孔(更换阀芯) (5)调压后锁紧调压螺母
振动和噪声大	(1)主阀芯在工作时径向力不平衡,导致溢流阀性能不稳定 (2)锥阀和阀座接触不好(圆度误差太大),导致锥阀受力不平衡,引起锥阀振动 (3)调压弹簧弯曲(或其轴线与端面不垂直),导致锥阀受力不平衡,引起锥阀振动 (4)通过流量超过公称流量,在溢流阀口处产生空穴现象 (5)通过溢流阀的溢流量太小,使溢流阀处于启闭临界状态而引起液压冲击	(1)检查阀体孔和主阀芯的精度,修换零件,过滤或换油 (2)封油面圆度误差控制在 0.005~0.01 mm 以内 (3)更换弹簧或修磨弹簧端面 (4)限在公称流量范围内使用 (5)控制正常工作的最小溢流量

任务实施

1. 画出设备上的压力控制阀的图形符号。
2. 能熟练调节压力控制阀对油路的压力进行调节。
3. 分组讨论,总结各种压力控制阀的应用场景。

任务 3　流量控制阀的认知

任务描述

了解常见流量控制阀的基本工作原理及符号,熟悉流量控制阀的种类及常见流量控制阀的不同应用场景。

相关知识

流量控制阀通过改变阀口通流面积来调节输出流量,从而控制执行元件的运动速度。常用的流量控制阀有节流阀和调速阀两种。

一、节流阀

图 3-3-1 为普通节流阀的结构原理图、图形符号和实物图。打开节流阀时,压力油从进油口 P_1 进入,经孔 a、阀芯 2 左端的轴向三角槽以及孔 b 和出油口 P_2 流出。阀芯在弹簧力的作用下始终紧贴在推杆 3 的端部。旋转手轮 4,可使推杆沿轴向移动,以改变节流口的通流面积,从而调节通过阀的流量。

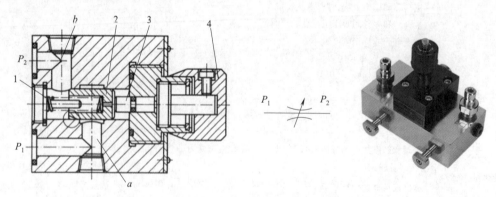

图 3-3-1　普通节流阀
1—弹簧;2—阀芯;3—推杆;4—手轮;a、b—孔

节流阀结构简单、体积小、使用方便、成本低,但负载和温度的变化对其流量稳定性的影

响较大,因此只适用于负载和温度变化不大或速度稳定性要求不高的液压系统。

二、调速阀

图 3-3-2 为调速阀工作原理图。从结构上来看,调速阀是在节流阀 2 前面串接一个定差减压阀 1 组合而成。液压泵的出口(即调速阀的进口)压力 p_1 由溢流阀调整基本不变,而调速阀的出口压力 p_3 则由液压缸负载 F 决定。油液先经减压阀产生一次压力降,将压力降到 p_2,p_2 经通道 e、f 作用到减压阀的 d 腔和 c 腔;节流阀的出口压力 p_3 又经反馈通道 a 作用到减压阀的上腔 b,当减压阀的阀芯在弹簧力 F_s、油液压力 p_2 和 p_3 作用下处于某一平衡位置时(忽略摩擦力和液动力等),则

$$p_2A_1+p_2A_2=p_3A+F_s \tag{3-1}$$

式中,A、A_1 和 A_2 分别为 b 腔、c 腔和 d 腔内压力油作用于阀芯的有效面积,且 $A=A_1+A_2$。

故

$$p_2-p_3=\Delta p=F_s/A \tag{3-2}$$

因为弹簧刚度较低,且工作过程中减压阀阀芯位移很小,可以认为 F_s 基本保持不变。故节流阀两端压力差 p_2-p_3 也基本保持不变,这就保证了通过节流阀的流量稳定。

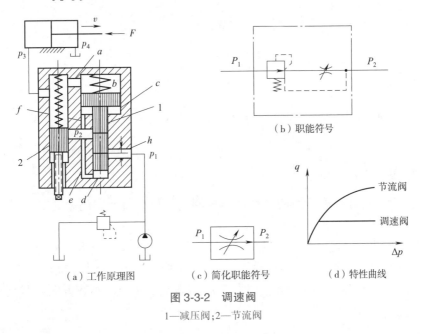

(a)工作原理图　　(c)简化职能符号　　(d)特性曲线

图 3-3-2　调速阀

1—减压阀;2—节流阀

三、流量控制阀的常见故障及其排除方法

流量控制阀的常见故障及其排除方法见表 3-3-1。

表 3-3-1　流量控制阀的常见故障及其排除方法

故障现象	原因分析	排除方法
无流量通过或流量极小	(1)节流口堵塞,阀芯卡住 (2)阀芯与阀孔配合间隙过大,泄漏大	(1)检查清洗,更换油液,提高油液清洁度,修复阀芯 (2)检查磨损、密封情况,修换阀芯
流量不稳定	(1)油中杂质黏附在节流口边缘上,通流截面减小,速度减慢 (2)节流阀内、外泄漏大,流量损失大,不能保证运行速度所需要的流量	(1)拆洗节流阀,清除污物,更换滤油器或油液 (2)检查阀芯与阀体之间的间隙及加工精度,修复或更换超差零件;检查有关连接部位的密封情况或更换密封件

任务实施

1. 画出设备上的流量控制阀的图形符号。
2. 能熟练调节流量控制阀对油路的流量进行调节。
3. 分组讨论,总结各种流量控制阀的不同应用场景。

任务 4　叠加阀的认知

任务描述

了解常见叠加阀的基本工作原理及符号,掌握叠加阀的种类,熟悉常见叠加阀的不同应用场景。

相关知识

叠加阀是近年获得迅速发展的一种液压控制阀。与普通液压阀相比,它具有许多显著优点。

叠加式液压阀简称叠加阀,其阀体本身既是元件,又是具有油路通道的连接体,阀体的上、下两面做成连接面。选择同一通径系列的叠加阀叠合在一起用螺栓紧固,即可组成所需的液压传动系统。叠加阀按功用不同,分为压力控制阀、流量控制阀和方向控制阀三类,其中方向控制阀仅有单向阀类,主换向阀不属于叠加阀。

一、叠加阀的结构和工作原理

叠加阀的工作原理与一般液压阀相同,只是具体结构有所不同。现以溢流阀为例,说明其结构和工作原理。

图 3-4-1(a)所示为先导型叠加式溢流阀的外形。其结构原理如图 3-4-1(b)所示,它由先导阀和主阀两部分组成,先导阀为锥阀,主阀相当于锥阀式的单向阀。其工作原理是:压力油

由 P 口进入主阀芯 6 右端的 e 腔,并经阀芯上阻尼孔 d 流至主阀芯左端 b 腔,再经小孔 a 作用于锥阀芯 3 上。当系统压力低于溢流阀调定压力时,锥阀关闭,主阀也关闭,阀不溢流;当系统压力达到溢流阀的调定压力时,锥阀芯 3 打开,b 腔的油液经锥阀口及孔 c 由油口 T 流回油箱,主阀芯右腔的油经阻尼孔 d 向左流动,于是使主阀芯的两端油液产生压力差。此压力差使主阀芯克服弹簧 5 的压力而左移,主阀口打开,实现了自油口 P 向油口 T 的溢流。调节弹簧 2 的预压缩量便可调节溢流阀的调整压力,即溢流压力。图 3-4-1(c)所示为其图形符号。

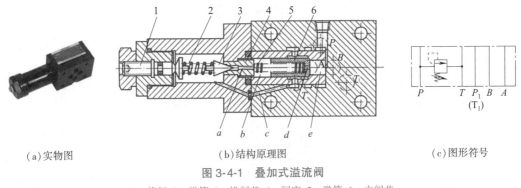

(a)实物图　　　(b)结构原理图　　　(c)图形符号

图 3-4-1　叠加式溢流阀

1—推杆;2—弹簧;3—锥阀芯;4—阀座;5—弹簧;6—主阀芯

二、叠加式液压系统的组装

叠加阀自成体系,每一种通径系列的叠加阀,其主油路通道以及螺钉孔的大小、位置、数量都与相应通径的板式换向阀相同。因此,将同一通径系列的叠加阀互相叠加,可直接连接而组成集成化液压系统。

图 3-4-2(a)所示为叠加式液压装置的外形。其结构如图 3-4-2(b)所示,最下面的是底板,底板上有进油孔、回油孔和通向液压执行元件的油孔,底板上面第一个元件一般是压力表开关,然后依次向上叠加各压力控制阀和流量控制阀,最上层为换向阀,用螺栓将它们固定成一个叠加阀组。一般一个叠加阀组控制一个执行元件。如果液压系统有几个需要集中控制的液压元件,则用多联底板,并排在上面组成相应的几个叠加阀组。

(a)两组和三组叠加阀组实物图　　　(b)结构示意图

图 3-4-2　叠加式液压装置

任务实施

1. 画出设备上的叠加阀的图形符号。
2. 能进行叠加阀的回路搭建。
3. 会分析叠加阀回路的故障,并进行简单的故障排除。

思考与练习

1. 液压控制元件有哪些?它们的基本原理是什么?
2. 在设备上完成下图的回路连接。
3. 小组讨论,总结各种叠加阀回路在不同应用场合中的功能。

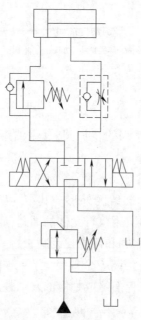

4. 对回路进行功能分析。该液压回路在设计时首先考虑该夹紧装置可以对不同材料的工件进行夹紧,由于工件材料的不同,其所需的夹紧力是不同的。在这个回路中钻头升降和工件夹紧共用一个液压泵供油,如采用溢流阀来调节夹紧压力,则会造成钻头钻孔时得不到足够的压力,所以在回路中设置一个溢流减压阀来调节夹紧压力。

这种用于降低回路中某一支路或某一执行元件工作压力的回路称为减压回路。在减压阀旁并联一个单向阀是为了减少液压缸活塞返回时的排油阻力,实现快速返回,同时也能延长减压阀的使用寿命。在夹紧过程中,为避免夹具对工件的损坏,夹紧速度应是可调的,所以回路中采用一个单向节流阀来对液压缸活塞的伸出速度进行调节。换向阀中位卸荷是为了保证液压缸尚未对工件进行夹紧前,钻头不会得到足够的压力产生误动作。如果将这个回路中的换向阀改为 M 形中位,那么在夹紧液压缸处于保压状态时,泵卸压、钻孔缸就无法获得足够的压力进行工作。

5. 分组讨论(3~5 人一组),对实验的数据和现象进行分析。在实验中可以观察到由于采用了回油节流,液压缸在伸出时左、右腔始终保持有较高压力,并在伸到位时左腔压力迅速上升到设定值,而右腔压力缓慢下降到零。对工件来说其夹紧力是逐渐升高的,这样对工件不易造成损伤。但这种压力变化过程不方便进行压力检测,甚至可能造成压力检测元件的误动作。

如要将回路改为电气控制,并在液压缸的左腔安装一个压力继电器对夹紧压力进行自动检测,那么该回路的节流方式就必须改为进油节流。因为采用进油节流液压缸在空载伸出时,左腔压力很低,在活动口碰上工件后压力缓慢上升到最高值,这就能保证压力继电器可靠动作,不会因误动作而对工件造成损伤。

项目 4
基本液压回路的安装与调试

任务 1　方向控制回路的安装与调试

任务描述

液压系统的换向回路有多种控制方式。本任务要求学生了解常用换向回路中的核心元器件,画出控制回路的电气原理图,并分析在不同控制条件下,液压缸是处于缩进还是浮动状态。

相关知识

一、二位四通电磁换向阀换向回路

运动部件的换向,一般可采用各种换向阀来实现。图 4-1-1 中就是利用了二位四通电磁换向阀来实现液压缸的左右换向。

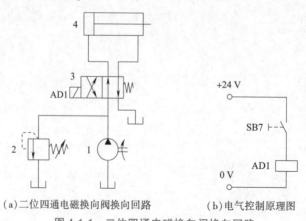

(a)二位四通电磁换向阀换向回路　　(b)电气控制原理图

图 4-1-1　二位四通电磁换向阀换向回路

1—液压泵;2—直动式溢流阀;3—二位四通电磁换向阀;4—双作用单活塞杆液压缸

当换向阀3处于图4-1-1(a)所示状态时,液压泵1压力油经换向阀3右位至液压缸4左腔,液压缸4活塞杆伸出。当按下SB7按钮,电磁铁AD1得电,液压泵1压力油经换向阀3左位至液压缸4右腔,液压缸4活塞杆退回。

二、三位四通电磁换向阀换向回路

如图4-1-2所示,AD2得电,液压泵1压力油经换向阀3右位至液压缸4左腔,液压缸4活塞杆伸出。当AD2失电,AD1得电,换向阀3处于中位,液压缸4停止运动。AD1得电,液压泵1压力油经换向阀3左位至液压缸4右腔,液压缸4活塞杆退回。

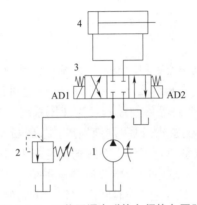

图4-1-2 三位四通电磁换向阀换向回路
1—液压泵;2—直动式溢流阀;3—三位四通电磁换向阀;4—双作用单活塞杆液压缸

电磁换向阀的换向回路应用最为广泛,尤其在自动化程度要求较高的组合机床液压系统中被普遍采用。这种换向回路曾多次出现于上面许多回路中,这里不再赘述。对于流量较大和换向平稳性要求较高的场合,电磁换向阀的换向回路已不能适应上述要求,往往采用手动换向阀或机动换向阀作先导阀,而以液动换向阀为主阀的换向回路。

任务实施

1. 通过网络查询图4-1-2中元件3的作用是什么。
2. 小组合作画出图4-1-2的电气控制原理图。
3. 当图4-1-2三位四通电磁换向阀换向回路中元件3不得电时,液压缸处于锁紧状态还是浮动状态?

任务2　压力控制回路的安装与调试

任务描述
分析常用的压力控制回路的方式,能够讲出常用压力控制回路的原理。

相关知识

一、调压及限压回路

压力控制回路是用压力阀来控制和调节液压系统主油路或某一支路的压力,以满足执行元件速度换接回路所需的力或力矩的要求。利用压力控制回路可实现对系统进行调压(稳压)、减压、增压、卸荷、保压与平衡等各种控制。当液压系统工作时,液压泵应向系统提供所需压力的液压油,同时,又能节省能源,减少油液发热,提高执行元件运动的平稳性。所以,应设置调压或限压回路。当液压泵一直工作在系统的调定压力时,就要通过溢流阀调节并稳定液压泵的工作压力。在变量泵系统中或旁路节流调速系统中用溢流阀(当安全阀用)限制系统的最高安全压力。当系统在不同的工作时间内需要有不同的工作压力时,可采用二级或多级调压回路。

1. 单级调压回路

如图4-2-1所示,通过液压泵1和溢流阀2的并联连接,即可组成单级调压回路。调节溢流阀的压力,可以改变泵的输出压力。当溢流阀的调定压力确定后,液压泵就在溢流阀的调定压力下工作。从而实现了对液压系统进行调压和稳压控制。

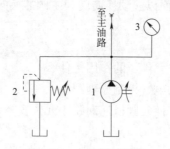

图 4-2-1　单级调压回路
1—液压泵;2—直动式溢流阀;3—耐震压力表

2. 二级调压回路

图4-2-2(a)所示为典型二级调压回路。该回路可实现两种不同的系统压力控制,由先导式溢流阀4和直动式溢流阀2各调一级。当二位三通电磁换向阀3处于图示位置时系统压

力由阀 4 调定。当阀 3 得电后换位时,系统压力由阀 2 调定,这时液压泵的溢流流量经主阀 4 回油箱,阀 2 亦处于工作状态,并有油液通过。

注意:阀 2 的调定压力一定要小于阀 4 的调定压力。

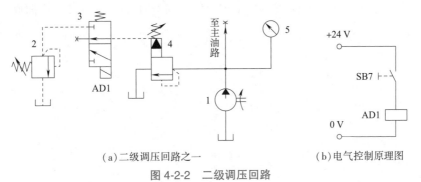

(a)二级调压回路之一　　　　(b)电气控制原理图

图 4-2-2　二级调压回路

1—液压泵;2—直动式溢流阀;3—二位三通电磁换向阀;4—先导式溢流阀;5—耐震压力表

图 4-2-3(a)为二级调压回路的另一例。活塞下降为工作行程,此时高压溢流阀 2 限制系统最高压力。活塞上升为非工作行程,用低压溢流阀 5 限制其最高压力。本回路常用于压力机的液压系统中。

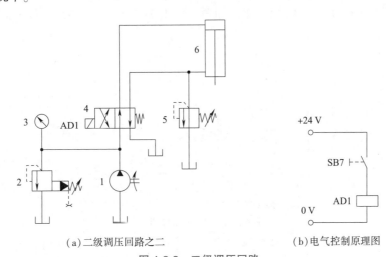

(a)二级调压回路之二　　　　(b)电气控制原理图

图 4-2-3　二级调压回路

1—液压泵;2—先导式溢流阀;3—耐震压力表;4—二位四通电磁换向阀;
5—直动式溢流阀;6—双作用单活塞杆液压缸

二、卸荷回路

执行元件工作间歇(或停止工作)时,不需要液压能,应自动将泵源排油直通油箱,组成卸荷回路,使液压泵处于无载荷运转状态,以便达到减少动力消耗和降低系统发热的目的。

1. 先导式溢流阀遥控口卸荷回路

图4-2-4(a)中,先导式溢流阀3的远程控制口直接与二位三通电磁阀4相连,便构成一种用先导式溢流阀遥控口卸荷回路。当换向阀4处于图示位置时,主油路系统压力由溢流阀3调节,当电磁铁AD1得电,换向阀4换向时,溢流阀3的远程控制口经换向阀4与油箱相通,系统压力卸荷。

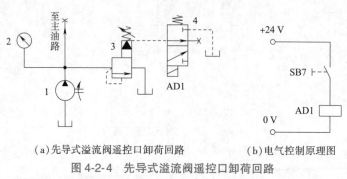

(a)先导式溢流阀遥控口卸荷回路　　(b)电气控制原理图

图 4-2-4　先导式溢流阀遥控口卸荷回路

1—液压泵;2—耐震压力表;3—先导式溢流阀;4—二位三通电磁换向阀

2. 换向阀卸荷回路

在图4-2-5(a)中,利用换向阀4使泵卸荷。但由于切换冲击压力大,该回路适用于低压小流量的系统。

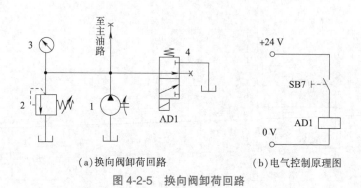

(a)换向阀卸荷回路　　(b)电气控制原理图

图 4-2-5　换向阀卸荷回路

1—液压泵;2—直动式溢流阀;3—耐震压力表;4—二位三通电磁换向阀

三、减压回路

当泵的输出压力是高压而局部回路或支路要求低压时,可以采用减压回路,如机床液压系统中的定位、夹紧、回路分度以及液压元件的控制油路等,它们往往要求比主油路较低的压力。减压回路较为简单,一般是在所需低压的支路上串接减压阀。采用减压回路虽能方便地获得某支路稳定的低压,但压力油经减压阀口时要产生压力损失。

图4-2-6(a)为一典型的减压回路,主要应用于夹紧回路中。液压泵供油最高压力由溢流

阀 2 根据主油路的需要来调节。主油路并联减压阀 3 后,使夹紧缸 7 获得较低的压力,其夹紧压力由减压阀 3 调节且可保持压力恒定。单向阀 4 起短时保压之用,当主油路压力低于减压阀调整压力时,防止夹紧力降低。

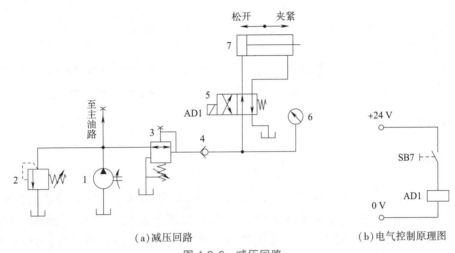

(a)减压回路　　　　　　　(b)电气控制原理图

图 4-2-6　减压回路

1—液压泵;2—直动式溢流阀;3—直动式减压阀;4—单向阀;
5—二位四通电磁换向阀;6—耐震压力表;7—双作用单活塞杆液压缸

任务实施

图 4-2-7 为二级调压回路的另外一例子,试分析实现二级调压的原理。

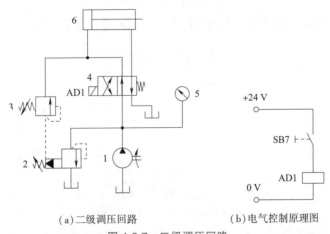

(a)二级调压回路　　　　　　(b)电气控制原理图

图 4-2-7　二级调压回路

1—液压泵;2—先导式溢流阀;3—直动式溢流阀;4—二位四通电磁换向阀;
5—耐震压力表;6—双作用单活塞杆液压缸

任务3 速度控制回路的安装与调试

任务描述

速度控制回路涉及液压系统的速度调节和变换问题。常用的速度控制回路有调速回路、快速回路等。要求学生掌握调速阀双向旁路节流调速回路的工作原理及特点。

相关知识

一、调速回路

从液压马达的工作原理可知,液压马达的转速 n_m 由输入流量和液压马达的排量 V_m 决定,即 $n_m = q/V_m$;液压缸的运动速度 v 由输入流量和液压缸的有效作用面积 A 决定,即 $v = q/A$。通过上面的关系可以知道,要想调节液压马达的转速 n_m 或液压缸的运动速度 v,可通过改变输入流量 q、改变液压马达的排量 V_m 和改变缸的有效作用面积 A 等方法来实现。由于液压缸的有效面积 A 是定值,只有改变流量 q 的大小来调速。而改变输入流量 q,可以通过采用流量阀或变量泵来实现。改变液压马达的排量 V_m,可通过采用变量液压马达来实现。以下是常用的调速回路。

1. 进油节流调速回路

液压缸的速度负载特性,液压缸在稳定工作时,其受力平衡方程式为

$$p_1 A = F + p_2 A$$

泵的供油压力 p_p 由溢流阀调定为恒定,故节流阀两端的压力差为

$$\Delta p = p_p - p_1 = p_p - \frac{F}{A}$$

节流阀进入液压缸的流量为

$$q_1 = K A_T \Delta p^m = K A_T \left(p_p - \frac{F}{A} \right)^m$$

速度负载特性方程为

$$v = \frac{q_1}{A} = \frac{K A_T}{A} \left(p_p - \frac{F}{A} \right)^m$$

上式为进油节流调速回路的速度负载特性方程。相同负载下工作时,这种回路调速范围较大。当调定后,速度随负载的增大而减小,故这种调速回路的速度负载特性较低。

由图4-3-1进油口节流调速的速度负载特性曲线可以看出,当节流阀开口 A_T 不变时,活塞的运动速度 v 随负载 F 的增加而

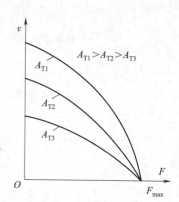

图4-3-1 进油口节流调速的速度负载特性曲线

降低,速度刚度较低。当节流阀开口 A_T 一定时,负载较小的区段曲线较平缓,速度刚度高,负载较大的区段曲线较陡,速度刚度较低。在相同负载下工作时,当节流阀开口较小,活塞的速度 v 较低,曲线较平缓,速度刚度高;节流阀开口较大,活塞的速度 v 较高,曲线较陡,速度刚度低。

图 4-3-2(a)所示为单向节流阀的进油节流调速回路,单向节流阀 4 位于执行机构 5 的进油路上,适用于以正载荷操作的液压缸。液压泵的余油经过溢流阀 2 排出,以溢流阀设定压力工作。这种回路效率低,调速范围大,适用于轻载低速工况。若将阀 4 改为单向调速阀,回路为调速阀进油节流调速回路,速度稳定性比用节流阀好。

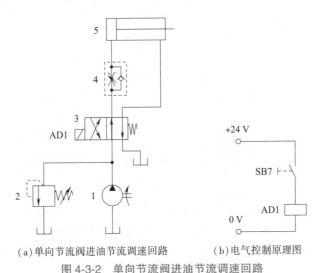

(a)单向节流阀进油节流调速回路　　(b)电气控制原理图

图 4-3-2　单向节流阀进油节流调速回路

1—液压泵;2—直动式溢流阀;3—二位四通电磁换向阀;4—单向节流阀;5—双作用单活塞杆液压缸

图 4-3-3(a)所示为调速阀双向进油节流调速回路,试分析其工作原理及特点。

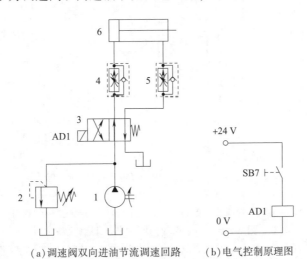

(a)调速阀双向进油节流调速回路　　(b)电气控制原理图

图 4-3-3　调速阀双向进油节流调速回路

1—液压泵;2—直动式溢流阀;3—二位四通电磁换向阀;4、5—调速阀;6—双作用单活塞杆液压缸

2. 回油节流调速回路

图 4-3-4(a)所示为单向节流阀的回油节流调速回路,单向节流阀 4 位于执行机构 5 的回油路上,适用于工作执行元件产生负载荷或载荷突然减少的情况。液压泵的输出压力为溢流阀的调定压力,与载荷无关,效率较低。它的优点是可产生背压,能抗拒负的载荷产生,防止突进。

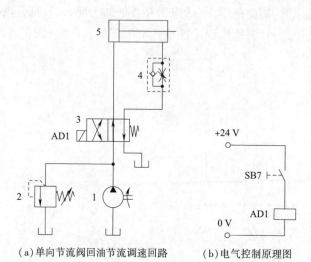

(a)单向节流阀回油节流调速回路　　(b)电气控制原理图

图 4-3-4　单向节流阀的回油节流调速回路

1—液压泵;2—直动式溢流阀;3—二位四通电磁换向阀;4—单向节流阀;5—双作用单活塞杆液压缸

图 4-3-5(a)所示为调速阀双向回油节流调速回路,试分析其工作原理及特点。

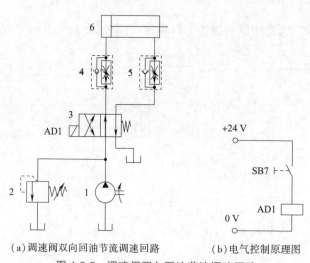

(a)调速阀双向回油节流调速回路　　(b)电气控制原理图

图 4-3-5　调速阀双向回油节流调速回路

1—液压泵;2—直动式溢流阀;3—二位四通电磁换向阀;4、5—调速阀;6—双作用单活塞杆液压缸

3. 旁路节流调速回路

图 4-3-6 所示为旁路节流调速回路负载特性曲线。当增大节流阀开度,活塞运动速度减小;当节流阀的开度不变时,负载增加活塞运动速度下降很快,其速度刚度比进、回油节流调速低。在负载一定时,节流阀的开度越小,其速度刚度越高,能承受的最大负载就越大。液压泵的工作压力随负载变化而变化,回路中只有节流损失而无溢流损失,因此这种回路的效率较高因液压缸的回油腔无背压力,所以其运动平稳性较低,不能承受负值负载。

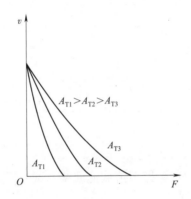

图 4-3-6　旁路节流调速回路负载特性曲线

图 4-3-7(a)所示为单向节流阀旁路节流调速回路。回路中余油直接由节流阀 4 排入油箱,液压泵的压力随载荷而变,其安全阀仅在油压超出安全压力时才打开,所以效率较高。若将阀 4 换为单向调速阀,回路即为调速阀旁路节流调速回路。

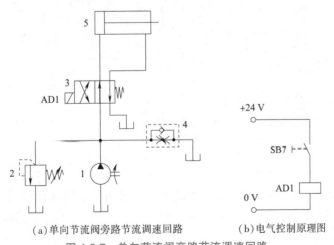

(a)单向节流阀旁路节流调速回路　　(b)电气控制原理图

图 4-3-7　单向节流阀旁路节流调速回路

1—液压泵;2—直动式溢流阀;3—二位四通电磁换向阀;4—单向节流阀;5—双作用单活塞杆液压缸

以上三种回路的优缺点比较见表4-3-1。

表4-3-1 调速回路的比较

回路类主要性能		节流调速回路			
		用节流阀		用调速阀	
		进回油	旁路	进回油	旁路
机械特性	速度稳定性	较差	差	好	
	承载能力	较好	较差	好	
调速范围		较大	小	较大	
功率特性	效率	低	较高	低	较高
	发热	大	较小	大	较小
适用范围		小功率、轻载的中、低压系统			

任务实施

1. 图4-3-8所示为调速阀双向旁路节流调速回路。试分析其工作原理及特点。

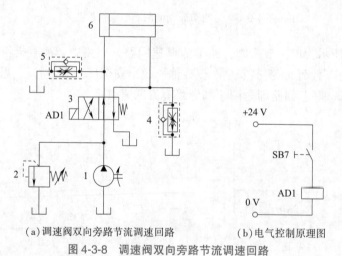

(a) 调速阀双向旁路节流调速回路　　(b) 电气控制原理图

图4-3-8　调速阀双向旁路节流调速回路

1—液压泵；2—直动式溢流阀；3—二位四通电磁换向阀；4、5—调速阀；6—双作用单活塞杆液压缸

2. 分析完成后，分组讨论这种双向旁油路节流调速回路与单向旁路节流调速回路的区别。

任务4 叠加阀控制回路的安装与调试

任务描述

了解叠加阀的优缺点、组装步骤及常用叠加阀回路的种类。叠加阀组中起控制作用的阀类型很多,每种类型的区别是什么?还有哪些其他阀组?

相关知识

一、叠加阀及系统的介绍

叠加阀是在集成块和油路板的基础上发展起来的新型元件,具有结构紧凑、体积小、维修简单等优点。用叠加阀组成的液压系统,不需要另外的连接件,以自身的阀体作为连接体,直接叠合而成,如图4-4-1所示。这种系统避免了管路、接头、法兰所带来的泄漏、振动、噪声等问题,使液压系统大为紧凑和简化。叠加阀与普通液压阀在工作原理上完全相同,但在具体结构和连接方式上有其特点,因而它自成系列。在系统中每个叠加阀,既起到控制元件的功能,又起通道体作用。同一种通径的叠加阀,包括各种压力阀、流量阀、方向阀等组成一个系列。在同一系列中的各种阀的油口位置、螺钉孔位置、连接尺寸都相同,只要按一定次序把各个阀叠加起来,便可组成各种典型的液压系统。

二、叠加阀的组装

叠加阀的组装要求与传统板式单向阀大体相同。但要特别留意各阀体的安装面位置是否配合对应,以避免出现泄漏。组装步骤如下:

(1)根据液压回路设计,先叠放叠加阀与电磁换向阀,将叠加阀带O形圈的一面朝向基础板。

(2)在叠加阀插进螺栓之前,确认叠加阀的油孔位置正确无误,对齐叠加阀螺钉安装孔。

(3)插进叠加阀专用安装螺钉,确保每个螺钉均用规定扭力拧紧。

图4-4-1所示为组装好的三组叠加阀组。

图4-4-1 三组叠加阀组

三、常用的叠加回路

1. 叠加式进油节流调速回路

图4-4-2所示为单向节流阀的进油节流调速回路,单向节流阀8位于执行机构7的进油

路上,适用于以正载荷操作的液压缸。这种回路效率虽然不高,但调速范围大,适用于轻载、低速工况。这种回路需要的油管数与前面板式阀搭建的进油口节流调速相比,大为减少。

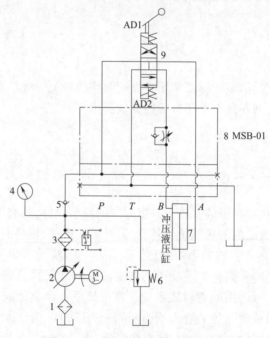

图 4-4-2　叠加式进油节流调速回路

1—吸油过滤器;2—变量叶片泵;3—高压滤油器;4—耐震压力表;5—单向阀;
6—直动式溢流阀;7—冲压液压缸;8—叠加式节流阀;9—三位四通换向阀(H 形)

在图 4-4-2 所示中用到的叠加式节流阀的实物图如图 4-4-3(a)所示,通过调节右侧旋钮就可以调节进油口的流量。从而可以调节液压缸伸出的速度。外加图 4-4-3(b)所示手动 H 形三位四通电磁换向阀换向。

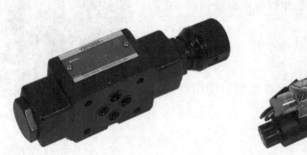
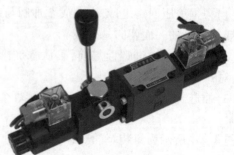

(a)MSB 型叠加式进油口节流阀　　　　(b)手动 H 形三位四通电磁换向阀

图 4-4-3　叠加阀组

2. 叠加式回油节流调速回路

图 4-4-4 所示为单向节流阀的回油节流调速回路,单向节流阀 8 位于执行机构 7 的回油路上,适用于工作执行元件产生负载荷或载荷突然减少的情况。液压泵的输出压力为溢流阀的调定压力,与载荷无关,效率较低。它的优点是可产生背压,能抗拒负的载荷产生,防止突进。

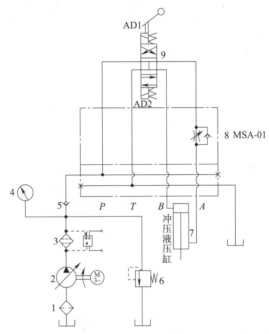

图 4-4-4　叠加式回油节流调速回路

1—吸油过滤器;2—变量叶片泵;3—高压滤油器;4—耐震压力表;5—单向阀;
6—直动式溢流阀;7—冲压液压缸;8—叠加式节流阀;9—三位四通换向阀(H 形)

图 4-4-4 中用到的叠加式节流阀的实物图如图 4-4-5(a)所示。通过调节左侧旋钮就可以调节回油口的流量,从而可以调节液压缸伸出的速度。另外,图 4-4-5(b)所示的手动 H 形三位四通电磁换向阀可用于换向。

(a) MSA 型叠加式回油口节流阀　　　　(b) 手动 H 形三位四通电磁换向阀

图 4-4-5　叠加阀组

3. 叠加式锁紧回路

锁紧回路的功能是使工作部件能在任意位置上停留,以及在停止工作时,防止在受力情况下发生移动。

图 4-4-6 中用到的叠加式液压锁的实物图如图 4-4-7(a)所示。

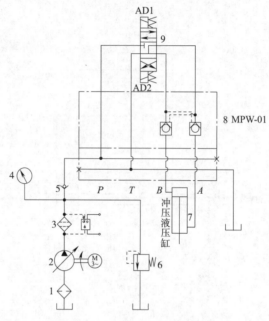

图 4-4-6　液控单向阀锁紧回路

1—吸油过滤器;2—变量叶片泵;3—高压滤油器;4—耐震压力表;5—单向阀;
6—直动式溢流阀;7—冲压液压缸;8—叠加式液压锁;9—三位四通换向阀(Y 形)

(a)叠加式液压锁　　　　　　(b)Y 形三维四通电磁换向阀

图 4-4-7　叠加阀组

对比图 4-4-8 所示的板式阀的液控锁回路连接与调试,会发现使用叠加式液控锁回路可以省去很多管路,而且在管路连接时不易出错。

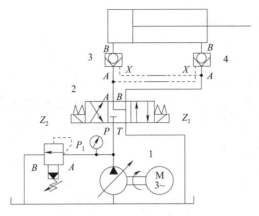

图 4-4-8 板式阀的液控锁回路图

任务实施

1. 在图 4-4-8 所示的液控锁回路中,为什么三位四通换向阀使用的是 Y 形的?如果换成 O 形的还能实现自锁吗?
2. 查阅资料,看看使用叠加式阀组有哪些缺点。
3. 除了叠加阀组外,是否还有其他类型的阀组,比如插装阀组。找出插装阀组的应用场景。

任务 5　比例阀控制回路的安装与调试

任务描述

比例控制阀是通过电液伺服系统的放大元件对输入的小功率电流信号进行转换,并以放大的液压功率(负载压力和负载流量)输出,实现对执行元件的位移、速度、加速度及力的控制的阀。常用比例控制阀有哪几种?安装时有哪些注意事项?

相关知识

一、比例控制阀的组成

比例控制阀通常由电气-机械转换装置、液压放大器和反馈机构三部分组成。电气-机械转换装置用来将输入的电信号转换为角位移或直线位移输出,输出角位移的装置称为力矩马达,输出直线位移的装置称为力马达。

液压放大器接受小功率的电气-机械转换装置输入的角位移或直线位移信号,对大功率的压力油进行调节和分配,实现控制功率的转换和放大。反馈和平衡机构是指电液伺服阀输出的流量或压力与输入电信号成比例的特性。

二、比例调速阀的连接与调试

1. 比例调速阀的连接

根据图 4-5-1 所示的比例调速阀控制液压原理图连接油路,并将直动式溢流阀 10 调压手柄逆时针旋至最大开度。根据图 4-5-2 所示的比例调速阀手动控制接线图接线,并将"输入设定"电位器旋钮逆时针旋转至零位。将泵站电控箱内手/自动切换二位旋钮开关打到【手动】,柱塞泵及叶片泵二位旋钮开关均打到【关】。

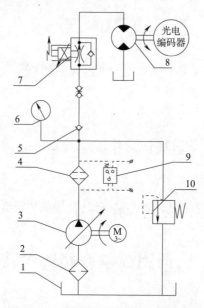

图 4-5-1　比例调速阀控制液压原理图

1—油箱;2—吸油过滤器;3—变量叶片泵;4—高压过滤器;5—单向阀;
6—压力表;7—比例调速阀;8—液压马达;9—压差发讯器;10—直动式溢流阀

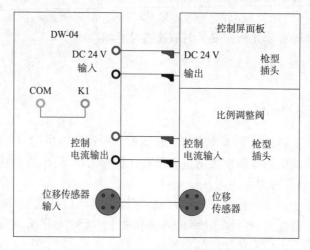

图 4-5-2　比例调速阀手动控制接线图

2. 比例调速阀手动控制实训

给设备上电,按下控制屏面板上的【启动】按钮,将泵站电控箱内叶片泵二位旋钮开关打到【开】,变量叶片泵启动,顺时针调节直动式溢流阀10手柄,调节系统压力至4 MPa。系统压力调定后,将其锁紧螺母锁紧。

将控制屏面板DC 24 V电源二位旋钮开关打到【开】,DW-04比例调速阀控制模块上的"反馈断路"指示灯不亮,"手动电压输入"指示灯亮,顺时针调节"输入设定"电位器旋钮,至一定圈数,液压马达转动,不断调节电位器旋钮,马达速度不断增大,逆时针调节电位器旋钮,马达速度不断减小。实训完成后,逆时针调节DW-04电位器旋钮至零位,拔下COM与K1之间的连接线,叶片泵、DC 24 V电源及AC 220 V电源二位旋钮开关打到【关】,按下控制屏面板上的【停止】按钮。

3. 基于PLC的比例调速阀PID控制实训

根据图4-5-3所示的西门子系统PLC接线图及图4-5-4所示的西门子系统比例调速阀接线图连接线路,检查无误后给设备上电,控制屏面板DC 24 V电源及AC 220 V电源二位旋钮开关打到【开】。

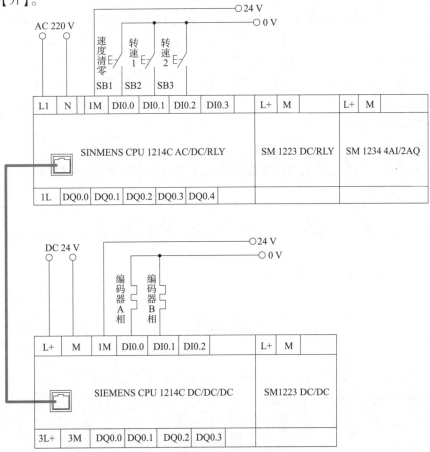

图4-5-3 西门子系统PLC接线图

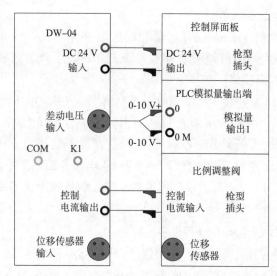

图 4-5-4　西门子系统比例调速阀接线图

给设备上电,按下控制屏面板上的【启动】按钮,将对应比例调速阀 PID 控制程序下载到相应 PLC 中。将泵站电控箱内叶片泵二位旋钮开关打到【开】,变量叶片泵启动。分别按下 SB2、SB3 按钮开关设定液压马达转速分别为 50 r/min,80 r/min,观察液压马达转速变化情况,若不理想可进行 PID 相关参数调节,SB1 按钮为转速清零。实训结束后,将叶片泵、DC 24 V 电源及 AC 220 V 电源二位旋钮开关打到【关】,按下控制屏面板上的【停止】按钮,拆下连接线,结束项目实训。

4. 注意事项

进行 PLC 程序下载时,插上和拔出连接线之前,务必将对应的 PLC 电源断开。

任务实施

1. 分组讨论。每 3~5 人分为一组,然后根据实训情况,讨论比例调速阀液压回路安装调试问题。
2. 分组讨论本次液压回路安装调试中的收获体会。

思考与练习

1. 在液压系统中,当工作部件停止运动后,使泵卸荷有什么好处?
2. 如何调节执行元件的运动速度?

项目 5

气压传动基础

任务 1 气压传动系统的工作原理及组成

任务描述

了解气压传动系统的工作原理及其组成,熟悉常用气压传动系统的主要应用场景。

相关知识

一、气压传动系统的工作原理

气压传动是以压缩空气为工作介质进行能量传递和信号传递的一门技术。气压传动系统的工作原理是利用空压机把电动机或其他原动机输出的机械能转换为空气的压力能,然后在控制元件的作用下,通过执行元件把压力能转换为直线运动或回转运动形式的机械能,从而完成各种动作,并对外做功。气动传动是以压缩空气为传动介质,空气可以方便获得,用过的气体也可方便排出,避免了液压传动中由于泄漏造成污染等一系列问题。目前,气动技术已经慢慢地从汽车、采矿、钢铁、机械等行业迅速发展到化工、轻工、军事等行业。气动控制系统因介质是空气,尤其适用于食品、无线电元器件和医药等生产。通过选定特殊的气体介质,气动传动系统也可用于易爆、易燃、易受到冲击的场合。

图 5-1-1 所示的气动机械手是某自动生产设备和生产线上的重要装置,它可以根据各种自动化设备的工作需要,按照预定的控制程序动作。因此,在机械加工、冲压、锻造、铸造、装配和热处理等生产过程中,它被广泛用来搬运工件,以减轻工人的劳动强度,也可实现自动取料、上料、卸料和自动换刀的功能。气动机械手是机械手的一种,它具有结构简单、自重小、动作迅速、操作稳定、性能可靠和节能等优点。

二、气压传动系统的组成

气压传动系统和液压传动系统类似,也由四部分组成,分别是:

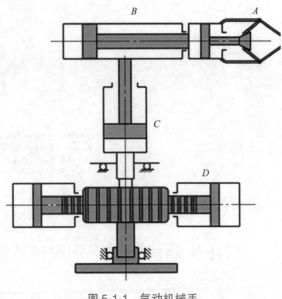

图 5-1-1　气动机械手

（1）气源装置。它是获得压缩空气的装置。其主体部分是空气压缩机,它将原动机供给的机械能转变为气体的压力能。

（2）控制元件。它是用来控制压缩空气的压力、流量和流动方向的装置,它使执行机构完成预定的工作循环。控制元件包括各种压力控制阀、流量控制阀和方向控制阀等。

（3）执行元件。它是将气体的压力能转换成机械能的一种能量转换装置。它包括实现直线往复运动的气缸和实现连续回转运动或摆动的气马达或摆动马达等。

（4）辅助元件。它是保证压缩空气的净化、元件的润滑、元件间的连接及消声等所必需的装置,包括过滤器、油雾器、管接头及消声器等。

任务实施

1. 通过网络查询哪些传动系统中使用了气压传动。举出三个实例。
2. 分组讨论。每3~5人分为一组,讨论气压传动的特点,并解释它与其他传动的差异。最后分组进行汇报。
3. 查阅我国气压传动系统的发展情况,客观评价,并提出自己的建议。

任务2　气压传动系统的特点

任务描述

气压传动系统的优缺点有哪些？请查阅资料,找出气压传动与其他传动方式的区别。

相关知识

一、气压传动的优点

气动技术广泛应用于机械、电子、轻工、纺织、食品、医药、包装、冶金、石化、航空、交通运输等各个工业部门。气动机械手、组合机床、加工中心、生产自动线、自动检测和实验装置等已大量涌现,它们在提高生产效率、自动化程度、产品质量、工作可靠性和实现特殊工艺等方面显示出极大的优越性。这主要是因为气压传动与机械、电气、液压传动相比有以下特点:

(1) 工作介质是空气,与液压油相比可节约资源,而且取之不尽、用之不竭。气体不易堵塞流动通道,用之后可将其随时排入大气中,不污染环境。

(2) 空气的特性受温度影响小。在高温下能可靠地工作,不会发生燃烧或爆炸。且温度变化时,对空气的黏度影响极小,故不会影响传动性能。

(3) 空气的黏度很小(约为液压油的万分之一),所以流动阻力小,在管道中流动的压力损失较小,便于集中供应和远距离输送。

(4) 相对液压传动而言,气动动作迅速、反应快,一般只需 0.02~0.3 s 就可达到工作压力和速度。液压油在管路中流动速度一般为 1~5 m/s,而气体的流速最小也大于 10 m/s,有时甚至达到音速,排气时还可达到超音速。

(5) 气体压力具有较强的自持能力,即使压缩机停机,关闭气阀,但装置中仍然可以维持一个稳定的压力。液压系统要保持压力,一般需要能源泵连续工作或另加蓄能器,而气体通过自身的膨胀性就可维持承载缸的压力不变。

(6) 气动元件可靠性高、寿命长。电气元件可运行百万次,而气动元件可运行 2 000 万~4 000 万次。

(7) 工作环境适应性好,特别是在易燃、易爆、多尘埃、强磁、辐射、振动等恶劣环境中,其传动和控制性能比液压、电子、电气系统优越。

(8) 气动装置结构简单,成本低,维护方便,过载能自动保护。

二、气压传动的缺点

(1) 由于空气的可压缩性较大,气动装置的动作稳定性较差,外负载变化时,对工作速度的影响较大。

(2) 由于工作压力低,气动装置的输出力或力矩受到限制。在结构尺寸相同的情况下,气压传动装置比液压传动装置输出的力要小得多。气压传动装置的输出力不宜大于 10~40 kN。

(3) 气动装置中的信号传动速度比光、电控制速度慢,所以不宜用于信号传递速度要求十分高的复杂线路中。同时实现生产过程的遥控也比较困难。但对一般的机械设备,气动信号的传递速度是能满足工作要求的。

(4) 噪声较大,尤其是在超音速排气时要加消声器。

气压传动与其他传动的性能比较见表 5-2-1。

表 5-2-1 气压传动与其他传动的性能比较

类型		操作力	动作快慢	环境要求	构造	负载变化影响	操作距离	无级调速	工作寿命	维护	价格
气压传动		中等	较快	适应性好	简单	较大	中距离	较好	长	一般	便宜
液压传动		最大	较慢	不怕振动	复杂	有一些	短距离	良好	一般	要求高	稍贵
电传动	电气	中等	快	要求高	稍复杂	几乎没有	远距离	良好	较短	要求较高	稍贵
	电子	最小	最快	要求特高	最复杂	没有	远距离	良好	短	要求更高	最贵
机械传动		较大	一般	一般	一般	没有	短距离	较困难	一般	简单	一般

任务实施

1. 通过网络查询常用气压传动系统的优缺点,并以现实生活中的案例说明该传动系统的优缺点。

2. 分组讨论。每 3~5 人分为一组,讨论气压传动的远距离传输问题,可以数控铣床中换刀装置为例。最后分组进行汇报。

任务 3 气压传动系统的应用

任务描述

气压传动系统具有清洁、快速的特点。那么它主要用在哪些方面呢?可用于汽车开闭门装置吗?气压传动系统的主要特点有哪些?它有哪些主要应用场景?

相关知识

气压传动简称气动,是指以压缩空气为工作介质来传递动力和控制信号。它是一门控制和驱动各种机械设备,实现生产过程机械化、自动化的技术。因为以压缩空气为工作介质具有防火、防爆、防电磁干扰、防振动、抗冲击、防辐射、无污染、结构简单、工作可靠等特点,所以,气动技术与液压、机械、电气、电子技术一起,互相补充,已发展成为实现生产过程自动化的一个重要手段,在机械工业、冶金工业、轻纺食品工业、化工、交通运输、航空航天、国防建设等各个行业已得到广泛的应用。

气动传动普遍应用于下列几个行业:

1. 机械制造业

目前,机械加工生产线的工件夹取及搬送,铸造生产线上的造型、捣固、合相,都已普遍采用了气动技术,汽车自动化生产线上的车体部件自动搬运与固定、自动焊接等工序中。图 5-3-1 所示为一个气动钻床的气压传动系统。

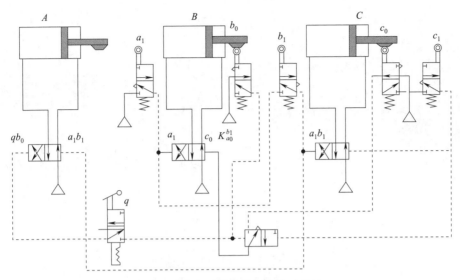

图 5-3-1 气动钻床的气压传动系统

2. 电子 IC 及电器行业

如用于硅片的搬运,元器件的插装与焊接,家用电器的组装等。

3. 石油化工行业

用管道输送介质的自动化流程绝大多数采用气动控制,如石油提炼加工、气体加工、化肥生产等。

4. 机器人行业

如装配机器人、喷漆机器人、搬运机器人以及爬墙、焊接机器人等。

5. 轻工食品包装行业

包括各种半自动或全自动包装生产线,例如酒类、油类、煤气罐装,各种食品的包装等。

6. 其他行业

如车辆刹车装置、车门开闭装置、颗粒物资的筛选、鱼雷导弹自动控制装置等。

任务实施

1. 通过网络查询"汽车门开闭装置的气动原理图"。

2. 分组讨论。每 3~5 人分为一组,然后就为什么气动技术在轻工业中应用最广泛和气动技术的原理是什么两个问题进行讨论,最后每组进行汇报。

思考与练习

1. 何为气压传动?气压传动的基本原理是什么?
2. 气压传动系统由哪几部分组成?各组成部分的主要内容是什么?
3. 简述气压传动的特点与应用。

项目 6
气动元件的认知

任务 1 气源装置

任务描述

气源装置和液压装置既很相似,又有差别。本任务要求了解常见气源装置的基本工作原理及符号,掌握气源装置的种类,熟悉常见气源装置在不同场合的应用。

相关知识

气源装置在气压传动系统中,为气动系统提供满足一定质量要求的压缩空气,是气压传动系统的重要组成部分。如图 6-1-1 所示,空气压缩机产生的压缩空气,经过降温、净化、减压、稳压等一系列操作后,才能供给控制元件和执行元件使用。而使用过后的压缩空气排向大气时会产生噪声。此时应采取相应措施,以降低噪声,改善劳动条件和环境质量。

图 6-1-1 空气压缩机

自然界中的空气是一种混合物,主要由氮、氧、水蒸气、其他微量气体和一些杂质组成。不同的环境和气候条件下,空气的组成成分也不同。气动系统工作时,压缩空气中水分和固体颗粒杂质等的含量是影响系统正常工作的重要因素。对空气进行压缩、净化,从而向各个

设备提供干净、干燥的压缩空气的装置称为压缩空气站或气源装置。图 6-1-2 所示为气源装置的工作示意图。

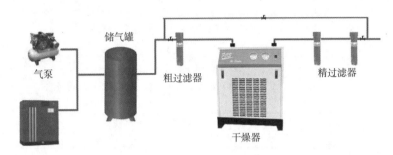

图 6-1-2　气源装置工作示意图

在对空气进行处理时，要先将空气压缩，然后再经过一系列过滤和干燥处理。图 6-1-3 所示为压缩空气站的详细工作流程。表 6-1-1 所示为气源装置各组成部分的符号及功能。

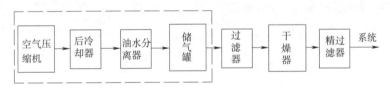

图 6-1-3　气源装置工作流程

表 6-1-1　气源装置组成部分

组成部分	职能符号		功能和作用
气泵			对空气进行压缩形成压缩空气
冷却器			将空压机出口的压缩空气冷却至 40 ℃以下，使得其中的大部分的水汽和变质油雾冷凝成液态水滴和油滴
油水分离器	手动		将经后冷却器降温析出的水滴和油滴等杂质从压缩空气中分离出
	自动		
储气罐			储存的压缩空并可以消除压力脉动，保证供气的连续性、稳定性
过滤器			进一步清除压缩空气中上油污、水和粉尘，以提高下游干燥器的工作效率，延长精过滤器的使用时间

续上表

组成部分	职能符号	功能和作用
干燥器	◇	进一步去除压缩空气中的水、油和灰尘
精过滤器	◇	再次对压缩空气中的油污、水和粉尘进行清除

二、气源调节装置

在实际操作中,从空气压缩站输出的压缩空气并不能满足气动元件对气源质量的要求,为使气源质量达到要求,常在气动系统前面安装气源调节装置。

1. 气源调节装置的组成

气源调节装置的组成如图6-1-4所示,由过滤器、减压阀和油雾器三部分组成,称之为三联件。

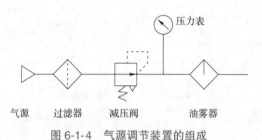

图 6-1-4　气源调节装置的组成

空气过滤器的作用是滤除压缩空气中所含的液态水滴、油滴、固体粉尘颗粒及其他杂质。空气过滤器一般由壳体和滤芯组成。按滤芯采用的材料不同,空气过滤器可分为纸质、织物、陶瓷、泡沫塑料和金属等形式。常用的是纸质式和金属式。

减压阀主要作用于将进气压力调节至系统所需的压力。

油雾器的作用是将润滑油雾化后喷入压缩空气管道的空气流,随空气进入系统中,润滑相对运动零件的表面。它有油雾型和微雾型两种。在气源压力大于 0.1 MPa 时,该油雾器允许在不关闭气路的情况下加油。供油量随气流大小而变化。油杯和视油帽采用透明材料制成,便于观察。油雾器要有良好的密封性、耐压性和滴油量调节性能。使用时,应参照有关标准合理调节起雾流量等参数,以达到最佳润滑效果。

由于一般气动系统的空气都是直接排入大气中,其所含的油雾对人体是有害的。一些特殊行业中不允许压缩空气中含有润滑油。随着技术的进步,一些新技术新工艺的应用,现在一些气动元器件已不需要在压缩空气中加润滑油。因此,气源调节装置就只有过滤器和减压阀,称之为二联件。

2. 气源调节装置的符号

图 6-1-5 所示为二联件和三联件的实物图及符号。一般我们在系统图中都用简化符号表示。

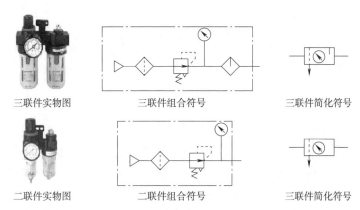

图 6-1-5　二联件和三联件的实物图及符号

三、空气压缩机的日常维护及保养事项

（1）保持机器清洁。

（2）储气罐的放水阀需每日打开一次以便排除油水。在湿气较重的地方，每 4 小时打开一次。

（3）检查润滑油的油位，每天一次，确保空压机的润滑。

（4）空气滤清器应 15 天清理或更换一次（滤芯为消耗品）。

（5）不定期检查各部位螺钉的紧固程度。

（6）润滑油最初运转 50 小时或一周后更换新油。以后每 300 小时换新油一次（使用环境较差者应 150 小时换一次油），每运转 36 小时加油一次。

（7）使用 500 小时（或半年）时，将气阀拆出清洗。

（8）每年将机器各部件清洁一次。

（9）定期检验全部防护罩、警告标志等安全防护装置。

（10）定期检查空压机的压力释放装置、停车保护装置及压力表（半年一次）、安全阀灵敏性，确保空压机处于正常工作状态。

（11）定期检查处在高温中的零部件，如阀、气缸盖、排气管道，清除附着在内壁上的油垢和积碳物。运转时，严禁触摸这些高温部件。

任务实施

1. 简述气动元件和液压元件的联系和区别，画出气源装置的符号。
2. 气动元件包括哪些组成部分，每个组成分别由什么元件组成。

任务 2 气动执行元件的认知

任务描述

了解常见气动执行元件的基本工作原理及符号,掌握气动执行元件的种类,熟悉常见气动执行元件在不同场合的应用。

相关知识

气动执行元件是将压缩空气的压力能转换为机械能的装置,包括气缸和气马达。气缸和气动马达的工作原理与液压缸和液压马达相似。气缸用于直线往复运动或摆动,气马达用于实现连续回转运动。

一、气缸

气缸是气动系统的执行元件之一。除几种特殊气缸外,普通气缸的种类及结构与液压缸基本相同。气缸常按以下方法分类:按作用方式分为单作用式和双作用式;按结构分为活塞式、柱塞式、叶片式、薄膜式;按功能分为普通气缸和特殊气缸(如冲击式、回转式和气-液阻尼式)。

1. 单作用气缸

图 6-2-1(a)所示为单作用气缸的实物。其结构原理如图 6-2-1(b)所示,压缩空气仅在气缸的一端进气并推动活塞(或柱塞)运动,而活塞(或柱塞)的返回是借助于其他外力,如弹簧力、重力等。单作用气缸多用于短行程及对活塞杆推力、运动速度要求不高的场合。

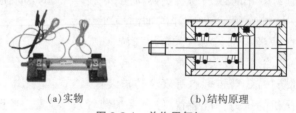

(a)实物　　　　　　(b)结构原理

图 6-2-1　单作用气缸

2. 薄膜式气缸

薄膜式气缸是一种利用压缩空气通过膜片推动活塞杆做往复直线运动的气缸。它由缸体、膜片、膜盘和活塞杆等主要零件组成。其功能类似于活塞式气缸,它分单作用式和双作用式两种,如图 6-2-2 所示。

薄膜式气缸的膜片可以做成盘形膜片和平膜片两种形式。膜片材料为夹织物橡胶、钢片或磷青铜片。常用的是夹织物橡胶,橡胶的厚度为 5~6 mm,有时也可用 1~3 mm。金属式膜片只用于行程较小的薄膜式气缸中。

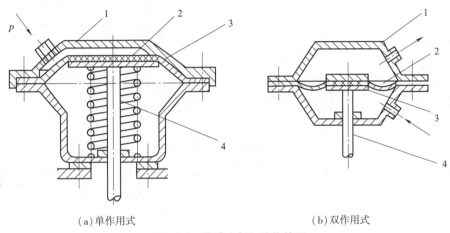

(a) 单作用式　　　　　(b) 双作用式

图 6-2-2　薄膜式气缸结构简图

1—缸体；2—膜片；3—膜盘；4—活塞杆

薄膜式气缸和活塞式气缸相比较，具有结构简单、紧凑、制造容易、成本低、维修方便、寿命长、泄漏小、效率高等优点。但是膜片的变形量有限，故其行程短（一般不超过 40～50 mm），且气缸活塞杆上的输出力随着行程的加大而减小。

3. 回转式气缸

图 6-2-3 所示为回转式气缸的结构原理。回转式气缸由导气头体、缸体、活塞、活塞杆等组成。这种气缸的缸体连同缸盖及导气头芯可被携带回转，活塞及活塞杆只能做往复直线运动，导气头体外接管路，固定不动。

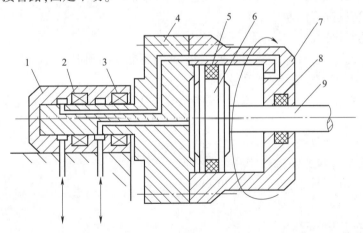

图 6-2-3　回转式气缸

1—导气头体；2、3—轴承；4—缸盖及导气头芯；5、8—密封装置；
6—活塞；7—缸体；9—活塞杆

在气动系统图中,气缸用相应的职能符号表示,常见气缸的职能符号见表 6-2-1。

表 6-2-1　常见气缸的职能符号

单作用气缸	双作用气缸		
	普通气缸	缓冲气缸	
弹簧压出	单活塞杆	不可调单向	可调单向
弹簧压入	双活塞杆	不可调双向	可调双向

二、气马达

气马达也是气动执行元件的一种。它的作用相当于电动机或液压马达,即输出力矩,拖动机构做旋转运动。

1. 气马达的分类及特点

气马达按结构形式可分为叶片式气马达、活塞式气马达和齿轮式气马达等。最为常见的是活塞式气马达和叶片式气马达。叶片式气马达制造简单,结构紧凑,但低速运动转矩小,低速性能不好,适用于中、低功率的机械,目前在矿山及风动工具中应用普遍。活塞式气马达在低速情况下有较大的输出功率,它的低速性能好,适宜于载荷较大和要求低速转矩的机械,如起重机、绞车、绞盘、拉管机等。

与液压马达相比,气马达具有以下特点:

(1) 工作安全。可以在易燃易爆场所工作,同时不受高温和振动的影响。

(2) 可以长时间满载工作而温升较小。

(3) 可以无级调速。控制进气流量,就能调节马达的转速和功率。额定转速以每分钟几十转到几十万转。

(4) 具有较高的启动力矩。可以直接带负载运动。

(5) 结构简单,操纵方便,维护容易,成本低。

(6) 输出功率相对较小,最大只有 20 kW。

(7) 耗气量大,效率低,噪声大。

2. 气马达的工作原理

图 6-2-4(a) 所示为叶片式气动马达的外形。其工作原理如图 6-2-4(b) 所示。压缩空气由 A 孔输入,小部分经定子两端密封盖的槽进入叶片底部,将叶片推出,使叶片紧贴在定子内壁上;大部分压缩空气进入相应的密封空间而作用在两个叶片上,由于两叶片长度不等,就产

生了转矩差,使叶片和转子按逆时针方向旋转。做功后的气体由定子上的 C 孔和 B 孔排出,若改变压缩空气的输入方向(即压缩空气由 B 孔进入,由 A 孔和 C 孔排出),就可改变转子的转向。

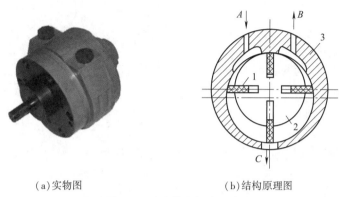

(a)实物图　　　　　　(b)结构原理图

图 6-2-4　叶片式气动马达

表 6-2-2 列出了各种气马达的特点及应用范围,可供选择和应用时参考。

表 6-2-2　各种气马达的特点及应用范围

形式	转矩	速度	功率	每千瓦耗气量 Q/ ($m^3 \cdot min^{-1}$)	特点及应用范围
叶片式	低转矩	高速度	由零点几千瓦到 1.3 kW	小型:1.8~2.3 大型:1.0~1.4	制造简单,结构紧凑,但低速启动转矩小,低速性能不好。适用于中低功率的机械,如手提工具、复合工具传送带、升降机、泵、拖拉机等
活塞式	中高转矩	低速或中速	由零点几千瓦到 1.7 kW	小型:1.9~2.3 大型:1.0~1.4	在低速时有较大的功率输出和较好的转矩特性。启动准确,且启动和停止特性均较叶片式好。适用于载荷较大和要求低速转矩较高的机械,如手提工具、起重机、绞车、绞盘、拉管机等
薄膜式	高转矩	低速度	小于 1 kW	1.2,u1.4	适用于要求精确控制、启动转矩极高和速度低的机械

任务实施

1. 画出设备上的气动执行元件图形符号。
2. 能熟练安装气动执行元件,且工作台面布局合理。
3. 小组讨论,总结各种气动执行元件的应用场景。

任务 3　气动控制元件的认知

任务描述

了解常见气动控制阀的基本工作原理及符号,掌握气动控制阀的种类,熟悉常见气动控制阀的不同应用场景。

相关知识

在气压传动系统中,气动控制元件指控制和调节压缩空气的压力、流量和方向的各种控制阀,其作用是保证气动执行元件(如气缸、气马达等)按设计的程序正常工作。气动控制元件按功能可分为压力控制阀、流量控制阀、方向控制阀以及能实现一定逻辑功能的逻辑元件。

一、压力控制阀

压力控制阀的作用是控制压缩空气的压力和依靠空气压力来控制执行元件的动作顺序。压力控制阀利用压缩空气作用在阀芯上的力和弹簧力相平衡的原理进行工作,主要有减压阀、溢流阀和顺序阀。

1. 减压阀(调压阀)

减压阀的作用是将出口压力调节为比进口压力低的调定值,并使输出压力保持稳定(又称调压阀)。减压阀分为直动式和先导式两种。

图 6-3-1(a)所示为 QTY 型直动式减压阀的结构。其工作原理是:当阀处于工作状态时,调节手柄 1、调压弹簧 2、3 及膜片 5,通过阀杆 6 使阀芯 8 下移,进气阀口被打开,有压气流从左端输入,经阀口节流减压后从右端输出。输出气流的一部分由阻尼孔 7 进入膜片气室,在膜片 5 的下方产生一个向上的推力,这个推力总是企图把阀口开度关小,使其输出压力下降。当作用于膜片上的推力与弹簧力相平衡后,减压阀的输出压力便保持一定。

当输入压力发生波动时,如输入压力瞬时升高,输出压力也随之升高,作用于膜片 5 上的气体推力也随之增大,破坏了原来的力的平衡,使膜片 5 向上移动,有少量气体经溢流口 4、排气孔 11 排出。在膜片上移的同时,因复位弹簧 10 的作用,使输出压力下降,直到新的平衡为止。重新平衡后的输出压力又基本上恢复至原值。反之,输出压力瞬时下降,膜片下移,进气口开度增大,节流作用减小,输出压力又基本上回升至原值。

调节手柄 1 使弹簧 2、3 恢复自由状态,输出压力降至零,阀芯 8 在复位弹簧 10 的作用下,关闭进气阀口。这样,减压阀便处于截止状态,无气流输出。

QTY 型直动式减压阀的调压范围为 0.05~0.63 MPa。为限制气体流过减压阀所造成的压力损失,规定气体通过阀内通道的流速在 15~25 m/s 范围内。

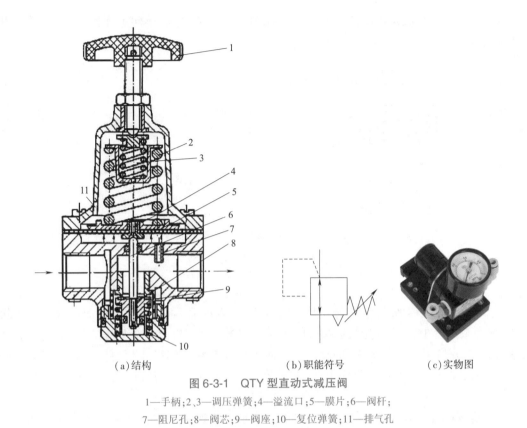

(a)结构　　　　　(b)职能符号　　　(c)实物图

图 6-3-1　QTY 型直动式减压阀

1—手柄;2、3—调压弹簧;4—溢流口;5—膜片;6—阀杆;
7—阻尼孔;8—阀芯;9—阀座;10—复位弹簧;11—排气孔

安装减压阀时,要按气流的方向和减压阀上所示的箭头方向,依照分水滤气器-减压阀油雾器的安装次序进行安装。调压时应由低向高调,直至规定的调压值为止。阀不用时应把手柄放松,以免膜片经常受压变形。

2. 溢流阀

溢流阀的作用是当系统中的压力超过调定值时,使部分压缩空气从排气口溢出,并在溢流过程中保持系统中的压力基本稳定,从而起过载保护作用(又称为安全阀)。溢流阀也分为直动式和先导式两种。按其结构可分为活塞式、膜片式和球阀式等。

直动式溢流阀的结构原理如图 6-3-2(a)所示,当输入压力超过调定值时,阀芯 3 便在下腔气压力作用下克服上面的弹簧力抬起,阀口开启,使部分气体排出,压力降低,从而起到过载保护作用。调节弹簧的预紧力可改变调定压力的大小。图 6-3-2(b)所示为其图形符号。图 6-3-2(c)所示为直动式溢流阀的外形。

二、流量控制阀

在气压传动系统中,有时需要控制气缸的运动速度,有时需要控制换向阀的切换时间和气动信号的传递速度,这些都需要调节压缩空气的流量来实现。流量控制阀就是通过改变阀的通流截面积来实现流量控制的元件。流量控制阀包括节流阀、单向节流阀、排气节流阀和快速排气阀等。

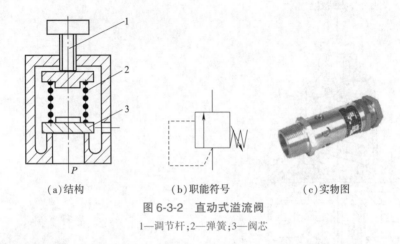

(a)结构　　　　(b)职能符号　　　(c)实物图

图 6-3-2　直动式溢流阀

1—调节杆；2—弹簧；3—阀芯

　　节流阀参考液压部分的节流阀以及单向节流阀，本章不再重复介绍，只简单介绍一下快速排气阀。图 6-3-3(a)所示为快速排气阀工作原理图。进气口 P 进入压缩空气，并将密封活塞迅速上推，开启阀口 2，同时关闭排气口 O，使进气口 P 和工作口 A 相通。当 P 口没有压缩空气进入时，在 A 口和 P 口压差作用下，密封活塞迅速下降，关闭 P 口，使 A 口通过 O 口快速排气。

　　快速排气阀常安装在换向阀和气缸之间。表示了快速排气阀在回路中的应用。它使气缸的排气不用通过换向阀而快速排出，从而加速了气缸往复的运动速度，缩短了工作周期。

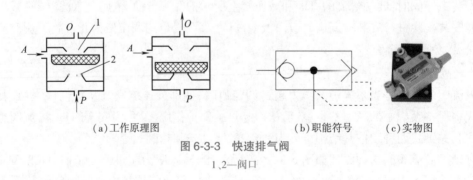

(a)工作原理图　　　　　　(b)职能符号　　(c)实物图

图 6-3-3　快速排气阀

1、2—阀口

三、方向控制阀

　　方向控制阀是气压传动系统中通过改变压缩空气的流动方向和气流的通断，来控制执行元件启动、停止及运动方向的气动元件。

　　根据方向控制阀的功能、控制方式、结构方式、阀内气流的方向及密封形式，可将方向控制阀分为几类，见表 6-3-1。

表 6-3-1　方向控制阀的分类

分类方式	形　式
按阀内气体的流动方向	单向阀、换向阀
按阀芯的结构形式	截止阀、滑阀
按阀的密封形式	硬质密封、软质密封
按阀的工作位数及通路数	二位三通、二位五通、三位五通等
按阀的控制操纵方式	气压控制、电磁控制、机械控制、手动控制

下面介绍几种典型的方向控制阀。

1. 单向型方向控制阀

单向型方向控制阀的作用是只允许气流向一个方向流动。它包括单向阀、梭阀和快速排气阀等。

(1) 单向阀。单向阀的结构原理如图 6-3-4(a)所示，当气流由 P 口进入时，气压力克服弹簧力和阀芯与阀体之间的摩擦力，使阀芯左移，阀口打开，气流正向通过。为保证气流稳定流动，P 腔与 A 腔应保持一定的压力差，使阀芯保持开启状态。当气流反向进入 A 腔时，阀口关闭，气流反向不通。图 6-3-4(b)所示为单向阀的图形符号。图 6-3-4(c)所示为单向阀的外形。

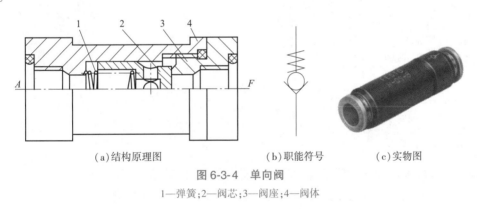

(a) 结构原理图　　(b) 职能符号　　(c) 实物图

图 6-3-4　单向阀

1—弹簧；2—阀芯；3—阀座；4—阀体

(2) 梭阀。梭阀的结构原理如图 6-3-5(a)所示，当需要两个输入口 P_1 和 P_2 均能与输出口 A 相通，而又不允许 P_1 和 P_2 相通时，就可以采用梭阀(或门)。当气流由 P_1 进入时，阀芯右移，使 P_1 与 A 相通，气流由 A 流出。与此同时，阀芯将 P_2 通路关闭。反之，P_2 与 A 相通，P_1 通路关闭。若 P_1 和 P_2 同时进气，则哪端压力高，A 就与那端相通，另一端自动关闭。图 6-3-5(b)所示为梭阀(或门)的图形符号。梭阀的应用很广，多用于手动与自动控制的并联回路中。图 6-3-5(c)所示为梭阀(或门)的外形。

(3) 快速排气阀。快速排气阀的结构原理如图 6-3-6(a)所示，当压缩空气进入进气口 P 时，使膜片 1 向下变形，打开 P 口与 A 口的通路，同时关闭排气口 O。当进气口 P 没有压缩空气进入时，在 A 口与 P 口压差的作用下，膜片向上复位，关闭 P 口，使 A 口通过 O 口快速排气。图6-3-6(b)所示为快速排气阀的图形符号。图6-3-6(c)所示为快速排气阀的外形。

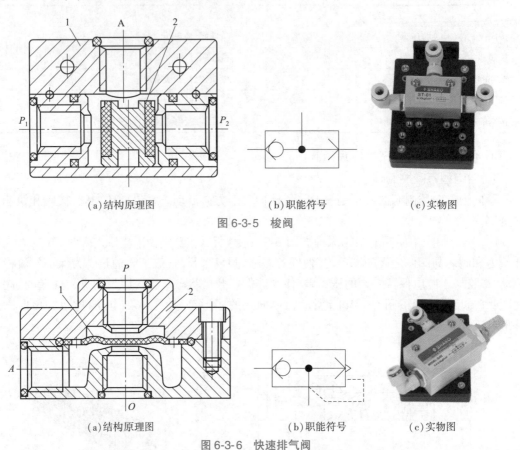

(a)结构原理图　　　　　(b)职能符号　　　　　(c)实物图

图6-3-5　梭阀

(a)结构原理图　　　　　(b)职能符号　　　　　(c)实物图

图6-3-6　快速排气阀

2. 换向型方向控制阀

换向型方向控制阀的作用是通过改变气流通道来改变气流方向,进而改变执行元件的运动方向。由于其换向原理与相同类型的液压换向阀相似,故不再重复。

四、气动逻辑元件

气动逻辑元件的作用是在系统中完成一定的逻辑功能。在输入信号的作用下,逻辑元件的输出信号状态只有"0"或"1"(表示"开"或"关"、"有"或"无"等)两种状态,属于开关元件(或数字元件)。它是以压缩空气为工作介质,利用元件内部的可动部件(如膜片、阀芯)在控制气压信号下动作,改变气流的输出状态,实现一定的逻辑功能。

气动逻辑元件种类很多,一般按下述方法分类:按工作压力分为高压型(0.2~0.8 MPa)、低压型(0.05~0.2 MPa)和微压型(0.005~0.05 MPa);按逻辑功能分为"是门"、"非门"、"或门"、"与门"和"双稳"元件等;按结构形式有截止式、膜片式和滑阀式等。表6-3-2列出了几种常用逻辑元件的图形符号及功用。

表 6-3-2 常用逻辑元件的图形符号及功用

类型	符号	功能
是门	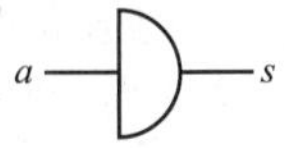 a ─▷─ s	元件的输入信号和输出信号之间始终保持相同的状态，即没有输入就没有输出，有输入才能输出
非门	 a ─▷○─ s	元件的输入信号和输出信号之间始终保持相反的状态，即有输入时无输出，而无输入时有输出
或门	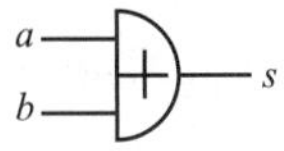 a,b ─▷─ s	有两个输入口和一个输出口，当一个口或两个口同时输入时，元件都有输出。两个输入口始终不通
与门	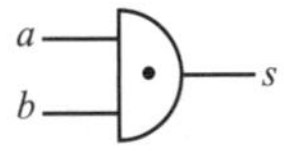 a,b ─▷·─ s	有两个输入口和一个输出口，只有两个输入口同时输入时才有输出
或非	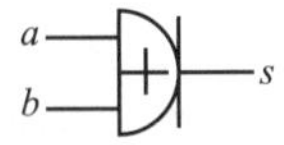 a,b ─▷○─ s	基本的两输入或非元件有两个输入口，当两个输入口都没有输入信号时，元件才有输出
禁门	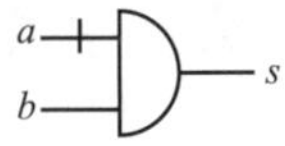 a,b ─▷─ s	只要有信号 a 存在，就禁止信号 b 输出；只有 a 不存在，才有 b 输出
双稳	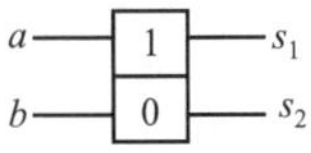 a─[1]─s_1 b─[0]─s_2	当输入信号 a 时，使 s_1 有输出，s_2 与排气口相通。a 信号消失，元件仍然保持 s_1 有输出状态。同样，输入信号 b 时，s_2 有输出，s_1 与排气口相通。b 信号消失，元件仍然保持 s_2 有输出状态。当两个输入同时进入时，元件状态取决于先输入的那个信号所对应的状态

五、气动辅助元件

气动辅助元件的功用是转换信号、传递信号、保护元件、连接元件以及改善系统的工况等。它的种类很多，主要有转换器、传感器、放大器、缓冲器、消声器、真空发生器和吸盘以及气路管件等。常用气动辅助元件的功用见表 6-3-3。

表 6-3-3 常用气动辅助元件的功用

类型		功能
转换器	气-液转换器	将压缩空气的压力能转换为油液的压力能，但压力值不变
	气-液增压器	将压缩空气的能量转换为油液的能量，但压力值增大，是将低压气体转换成高压油输出至负载液压缸或其他装置以获得更大驱动力的装置
	压力继电器	在气动系统中气压超过或低于给定压力(或压差)时发出电信号。另外，气-电转换器也是将气压信号转换为电信号的元件，其结构与压力继电器相似。不同的是压力不可调，只显示压力的有无，且结构较简单

续上表

类 型	功 能
传感器和放大器	气动位置传感器:将位置信号转换成气压信号(气测式)或电信号(电测式),进行检测气动放大器:气测式传感器输出的信号一般较小,在实际使用时,一般与放大器配合,以放大信号(压力或流量)
缓冲器	当物体运动时,由于惯性作用,在行程末端产生冲击。设置缓冲器可减小冲击,保证系统平稳、安全地工作
消声器	在气动元件的排气口安装消声器可降低排气的噪声,有的消声器还能分离和除去排气中的污染物
真空发生器和吸盘	真空发生器利用压缩空气的高速运动,形成负压而产生真空。真空吸盘利用其内部的负压将工件吸住。它们普遍用于薄板、易碎物体等的搬运

任务实施

1. 画出设备上的气动元件的图形符号。
2. 能熟练调节各种气动元件。
3. 会分析气动元件的故障,并进行简单的故障排除。

思考与练习

1. 液压执行元件有哪些？它们的基本原理是什么？
2. 液压执行元件有哪些？它们的基本原理是什么？
3. 小组讨论,总结气动控制元件的应用场景。

项目 7

气动回路

任务　气动基本回路的安装与调试

任务描述

了解常用的基本回路有哪些；能够画出常用的基本回路及计数回路的工作原理图；能够对这些回路进行分析。

相关知识

一、换向回路

1. 单作用缸的换向回路

图 7-1-1 所示为单作用气缸的换向回路。这是一种比较简单的换向回路，由一个两位三通换向阀来控制。按下按钮，单作用气缸伸出；松开按钮，单作用气缸在弹簧的作用下自动缩回。

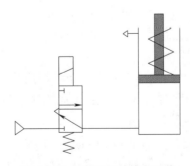

图 7-1-1　单作用缸的换向回路

2. 双作用气缸的换向回路

图 7-1-2 所示为各种双作用气缸的换向回路。图 7-1-2(a) 所示为比较简单的换向回路；图 7-1-2(b) 中还有中停位置，但中停定位精度不高；图 7-1-2(b)、(d)、(e) 中的两端控制电磁铁线圈或按钮不能同时操作，否则将出现误动作，其回路相当于双稳的逻辑功能；对图 7-1-2(f) 所示的回路，当 A 处有压缩空气时，气缸推出，反之，气缸退回。

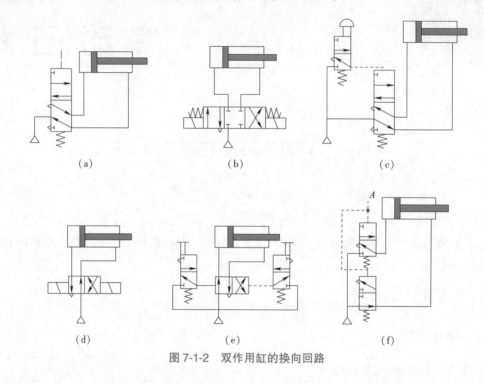

图 7-1-2 双作用缸的换向回路

二、速度控制回路

1. 单作用气缸的速度控制回路

图 7-1-3 所示为单作用气缸的速度控制回路。图 7-1-3(a) 中由两个方向相反的节流阀来控制气缸速度。当气缸伸出时，左边第一个节流阀起到调速作用；等气缸缩回时，第二个节流阀起到对气缸缩回起到节流调速作用。

图 7-1-3(b) 中由节流阀以及快速排气阀来控制单作用缸的速度，当气缸伸出时，节流阀起到节流调速的作用；当气缸缩回时，通过快速排气阀快速缩回。

2. 双作用气缸的速度控制回路

在气动系统中，对水平安装的气缸较少使用进口节流调速，主要是因为气缸运动时容易发生"爬行"或"跑空"现象。为获得稳定的运动速度，气动系统多采用出口节流调速，如图 7-1-4 所示。

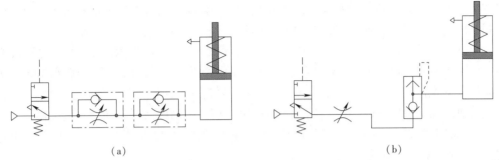

图 7-1-3　单作用缸的速度控制回路

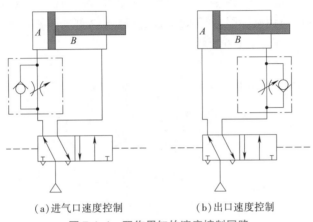

(a) 进气口速度控制　　(b) 出口速度控制

图 7-1-4　双作用缸的速度控制回路

3. 手动阀控制双向速度调节回路

当按下阀 3 时，节流阀 1 起到进气口速度调节；当气缸返回时，节流阀 2 起到节流调速作用，如图 7-1-5 所示。

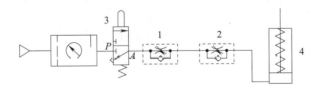

图 7-1-5　手动阀控制双向速度调节回路

三、压力控制回路

1. 一次压力控制回路

在一次压力控制回路中，空气压缩机由电动机驱动，当启动电动机后，空气压缩机产生的压缩空气经单向阀进入储气罐，储气罐内的压力上升，电接点式压力表显示压力值。当储气罐内的压力值上升到气压传动系统的最大限定值时,电接点式压力表内的指针碰到上触点,即控制其内的中间继电器断电,使电动机停止转动,空气压缩机也停止运转。储气罐内的压力不再上升。当储气罐内的压力值下降到气压传动系统的最小限定值时,电接点式压力表内的指针碰到下触点,即中间继电器闭合通电,使电动机重新开始转动,空气压缩机也重新开始运转向储气罐供压缩空气。电接点式压力表内的指针上下触点范围可调整。当电接点式压力表发生故障而失灵时,安全阀开启溢流,使气压传动系统的压力稳定在调定范围内。

特点:采用溢流阀结构简单,工作可靠,但气量浪费大;电接点压力表对电机及控制要求高,常用于对小型空压机的控制。

2. 二次压力控制回路

这种回路主要用于气源压力控制。它是由气动三大件——空气过滤器(分水滤气器)、减压阀与油雾器组成的压力控制回路。它采用溢流式减压阀对气源实行定压控制,是气动设备中必不可少的常用回路,如图 7-1-6 所示。

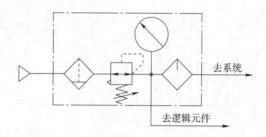

图 7-1-6 二次压力控制回路

3. 高低压力转换控制回路

这种回路由减压阀控制,同时输出一个高压力 p_1 和一个低压力 p_2。减压阀调节的压力不同,从而使两个压力也有所不同,如图 7-1-7 所示。

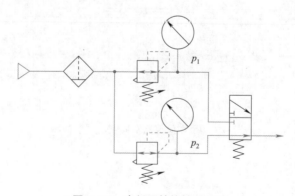

图 7-1-7 高低压转换控制回路

四、气缸同步动作回路

1. 双缸并联同步动作回路

当二位五通电磁阀 1 失电时,两个气缸同时伸出;当二位五通电磁阀 1 得电时,两个气缸缩回,如图 7-1-8 所示。

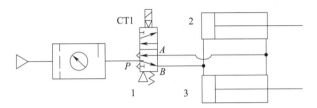

图 7-1-8 双缸并联同步动作回路

2. 双缸串联同步动作回路

当二位五通电磁阀 1 失电时,两个气缸同时伸出;当二位五通电磁阀 1 得电时,两个气缸缩回,如图 7-1-9 所示。

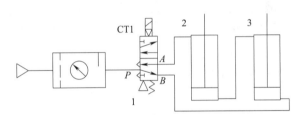

图 7-1-9 双缸串联同步动作回路

五、计数回路

在图 7-1-10 所示回路中,使阀 1 换位,则气管信号经阀 2 至阀 4 的左或右控制端使气缸推出或退回。阀 3 的换向位置,取决于阀 2 的搁置,而阀 2 的换位又取决于阀 4 和阀 5 的搁置。如图 7-1-10 所示,当阀 1 换位时,气信号经阀 2 至阀 3 的右端,气缸后退,同时阀 5 换位至左位;当阀 1 复位后,原通入阀 3 右控制端的气信号经阀 1 排空,阀 5 复位,于是气缸无杆腔的气经阀 5 至阀 2 左端,使阀 2 换至左位等待阀 1 的下一次信号输入。当阀 1 第二次按下后,气信号经阀 2 的左位至阀 4 左控制端使阀 4 换至左位,气缸前进,同时阀 4 将气路切断。待阀 1 复位后,阀 3 右控制信号经阀 2、阀 1 排空,阀 4 复位并将气导至阀 2 左端使其换至右位,又等待阀 1 下一次信号输入。这样,第 1,3,5,…次(奇数)按压阀 1,则气缸退回;第 2,4,6,…次(偶数)按压阀 1,则使气缸前进。

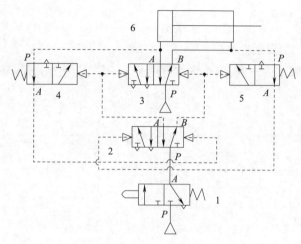

图 7-1-10 计数回路

任务实施

1. 通过学习,掌握全部气动基本回路。写出气动基本回路的归纳总结。

2. 分组讨论。每 3~5 人分为一组,每组分工,各画出一个常用气动回路图,最后分每组进行展示。

3. 在小组内相互讲解计数回路,并默画出计数气动回路的气动图。

思考与练习

1. 气动常用回路有哪些?

2. 小组讨论,总结各种气动常用回路的特点。

Task implementation

1   Understand all basic pneumatic circuits through the study, and make a summary.

2   Grouping and discussion: Group in 3-5 persons, draw the diagrams of commonly used pneumatic circuits, and exhibit the same by groups.

3   For counting circuit, interpret to each other by groups, and draw the pneumatic diagram of pneumatic counting circuit in mind.

Thinking and Exercise

1   What are the common pneumatic circuits?

2   Grouping and discussion: The characteristics of common pneumatic circuits are summarized.

V Counting Circuit

In the circuit shown in Figure 7-1-10, if valve 1 changes position, the air pipe signal enables to extend or retract the cylinder via the left or right control ends of valves 2 to 4. Commutation position of valve 3 depends on the placement of valve 2, and the commutation of valve 2 further depends on the placement of valves 4 and 5. As shown in Figure 7-1-10, when valve 1 changes position, air signal enables the cylinder to retract via the right ends of valves 2 to 3, and valve 5 changes to left position at the same time; when valve 1 is restored, the air signal that was formerly led to the right control end of valve 3 is discharged via valve 1, and valve 5 is restored, therefore, the air inside the cylinder chamber without rod enables vale 2 to change to left position via the left ends of valves 5 to 2, waiting for the next signal input from valve 1. When valve 1 is pressed again, the air signal enables valve 4 to change to left position via the left position of valve 2 to the left control end of valve 4, then cylinder moves forward and the air circuit of valve 4 is cut off at the same time. When valve 1 is restored, right control signal of valve 3 is discharged via valves 2 and 1, valve 4 is restored and leads the air to the left end of valve 2 to change it to the right position, waiting for the next signal input from valve 1 again. In this way, when pressing valve 1 at the 1st, 3rd, 5th,... (odd number) time, the cylinder retracts; when pressing valve 1 at the 2nd, 4th, 6th,... (even number) time, the cylinder moves forward.

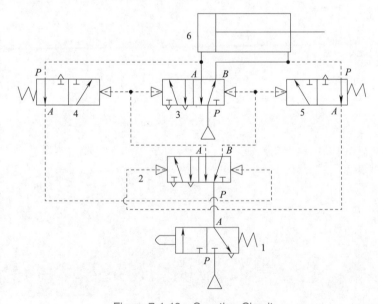

Figure 7-1-10 Counting Circuit

different pressure regulated by the relief valve results in different high and low pressures, see Figure 7-1-7.

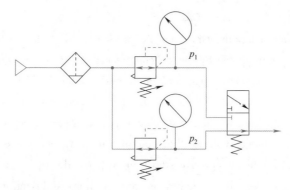

Figure 7-1-7 High- and Low-pressure Conversion Control Circuit

IV Cylinder Synchronous Action Circuit

1 Double-cylinder Parallel Synchronous Action Circuit

When the two-position five-way solenoid valve 1 is de-energized, the two cylinders extend simultaneously; and the two-position five-way solenoid valve 1 is energized, the cylinders retract, see Figure 7-1-8.

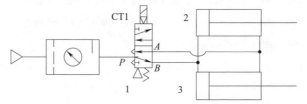

Figure 7-1-8 Double-cylinder Parallel Synchronous Action Circuit

2 Double-cylinder Series Synchronous Action Circuit

When the two-position five-way solenoid valve 1 is de-energized, the two cylinders extend simultaneously; and the two-position five-way solenoid valve 1 is energized, the cylinders retract, see Figure 7-1-9.

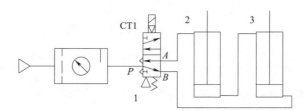

Figure 7-1-9 Double-cylinder Series Synchronous Action Circuit

III Pressure Control Circuits

1 Primary pressure control circuit

Primary pressure control circuit, where air compressor is driven by a motor, after the motor is started, the compressed air produced by the air compressor enters the air storage tank via a check valve, the pressure inside the air storage tank increases, and the electric contact type pressure gauge displays the pressure value. When the pressure inside air storage tankincrease to the maximum limit of pneumatic transmission system, the pointer inside electric contact type pressure gauge touches the upper contact, thereby the intermediate relay inside the pressure gauge is de-energized, the motor and the air compressor stop running, and the pressure inside air storage tank stops increasing. When the pressure inside air storage tank drops to the minimum limit of pneumatic transmission system, the pointer inside electric contact type pressure gauge touches the lower contact, thereby the intermediate relay is energized, the motor and the air compressor restart running and supplying air into the air storage tank; and the range between the upper and the lower contacts inside the electric contact type pressure gauge is adjustable. When the electric contact type pressure gauge fails, the safety valve is started for relieving pressure, so as to maintain the pressure of pneumatic transmission system within the set range.

Features: The overflow valve adopted has simple structure and reliable operation, but it wastes a large amount of air; the electric contact type pressure gauge has high requirements on the motor and control, and it is commonly used for the control of small size air compressor.

2 Secondary pressure control circuit

Mainly for pressure control of air supply: The pressure control circuit consisting of the three main pneumatic components-air filter (water-separating gas filter), relief valve and atomized lubricator. Overflow type relief valve is used for constant pressure control of air supply. It is an indispensable and commonly used circuit in pneumatic equipment, see Figure 7-1-6.

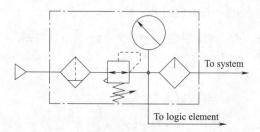

Figure 7-1-6 Secondary Pressure Control Circuit

3 High- and Low-pressure Conversion Control Circuit

A relief valve controls and simultaneously outputs the high and low pressures p_1 and p_2;

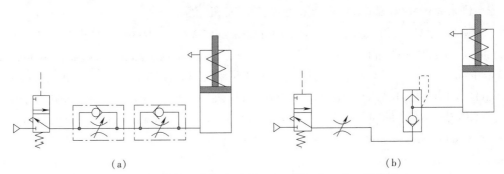

Figure 7-1-3 Speed Control Circuit of Single-acting Cylinder

2 Speed Control Circuit of Double-acting Cylinder

For a horizontally installed cylinder in pneumatic system, inlet throttling speed control is rarely used, this is mainly because the "creeping" or "air leaking" phenomenon occurs easily during movement of air cylinder. Outlet throttling speed control is mostly used for pneumatic system, in order to obtain stable movement speed, see Figure 7-1-4.

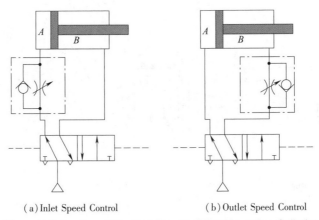

(a) Inlet Speed Control　　　　　　(b) Outlet Speed Control

Figure 7-1-4 Speed Control Circuit of Double-acting Cylinder

3 Bidirectional Speed Regulating Circuit of Manual Valve Control

When valve 3 is pressed, the throttle valve 1 serves as air inlet speed control; when the cylinder retracts, the throttle valve 2 serves as throttling speed control, see Figure 7-1-5.

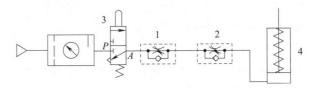

Figure 7-1-5 Bidirectional Speed Regulating Circuit of Manual Valve Control

2 Commutation Circuit of Double-acting Cylinder

Figure 7-1-2 shows the commutation circuit of double-acting cylinder. Figure 7-1-2(a) shows a simple commutation circuit; Figure 7-1-2(b) further comprises an intermediate stop position, but the intermediate stop positioning accuracy is not high; the control electromagnet coils or buttons at both ends of Figure 7-1-2(b), Figure 7-1-2(d), Figure 7-1-2(e) cannot be operated at the same time, otherwise, malfunction occurs, and the circuit is equivalent to the bistable logic function; for the circuit shown in Figure 7-1-2(f), when there is compressed air at point A, the cylinder extends, otherwise, it retracts.

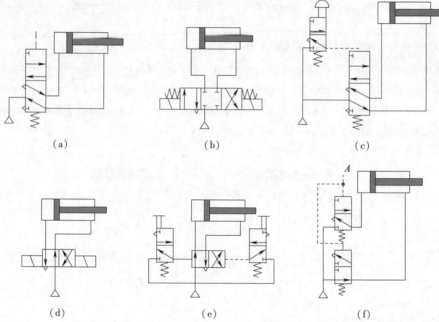

Figure 7-1-2 Commutation Circuit of Double-acting Cylinder

II Speed Control Circuits

1 Speed Control Circuit of Single-acting Cylinder

Figure 7-1-3 shows the speed control circuit of single-acting cylinder. Figure 7-1-3(a) shows the control of cylinder speed by two throttle valves of reversed directions; when the cylinder extends, the first throttle valve on the left regulates the speed while the cylinder is extending; and when the cylinder retracts, the second throttle valve serves as throttling speed control for retracting the cylinder.

Figure 7-1-3(b) shows the control of single-acting cylinder by throttle valve and quick exhaust valve; when the cylinder extends, the throttle vale serves as throttling speed control; and when the cylinder retracts, it retracts quickly by the quick exhaust valve.

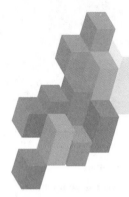

Part 7

Pneumatic Circuits

Task   Installation and Commissioning of Basic Pneumatic Circuits

Task description

Know the commonly used basic circuits, be able to draw schematic diagrams of the commonly used basic circuits and counting circuits, and also be able to analyze these circuits.

Relevant knowledge

I Commutation Circuits

1 Commutation Circuit of Single-acting Cylinder

Figure 7-1-1 shows the commutation circuit of single-acting cylinder. It is a relatively simple commutation circuit that is controlled by a two-position three-way reversing valve, where the single-acting cylinder extends after pressed the button and retracts automatically under the effect of spring after released the button.

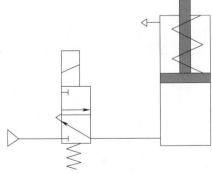

Figure 7-1-1   Commutation Circuit of Single-acting Cylinder

2 Be acquainted with adjustments of various pneumatic components.
3 Know how to analyze faults of pneumatic components, and to eliminate simple faults.

Thinking and Exercise

1 What are the hydraulic actuators? What about their basic working principles?
2 What are the hydraulic actuators? What about their basic working principles?
3 Discuss by groups, summarize the application scenarios of pneumatic control component.

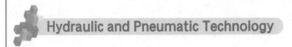

V Pneumatic Auxiliary Components

Pneumatic auxiliary components are used for signal conversion, signal delivery, element protection, element connection and system working condition improvement, etc. There aremany kinds of pneumatic auxiliary components, mainly including converter, sensor, amplifier, buffer, muffler, vacuum generator, suction disc and air pipe fittings, etc. Functions of commonly used pneumatic auxiliary components are shown in Table 6-3-3.

Table 6-3-3 Functions of Commonly Used Pneumatic Auxiliary Components

Type		Function
Converter	Pneumatic-hydraulic converter	Convert the pressure energy of compressed air into pressure energy of oil fluid, without changing the pressure value
	Pneumatic-hydraulic converter booster	Convert the energy of compressed air into energy of oil fluid, but increasing the pressure value. It is the device which converts the low-pressure air into high-pressure fluid, and outputs the high-pressure fluid to a load hydraulic cylinder or other device to obtain larger driving force
	Pressure relay	It gives electric signal when the air pressure in pneumatic system is higher or lower than the set pressure (or differential pressure). In addition, pneumatic-electrical converter is also a component that converts pressure signal into electrical signal; and its structure is similar to the pressure relay. However, its pressure is not adjustable; it indicates only whether there is pressure or not; and its structure simple
Sensor and amplifier		Pneumatic position sensor: convert position signal into pressure signal (gas logging) or electrical signal (electrical logging) to detect pneumatic amplifier; in general, the output signal of gas logging sensor is small; and in actual use, it is always used together with amplifier, so as to amplify the signal (pressure or flow)
Buffer		When an object is moving, a shock will be produced at the end of stroke due to inertia, so a buffer can be provided to reduce the shock, thereby ensuring the stable and safe work of system
Muffler		Installing a muffler at the exhaust vent of pneumatic component can reduce the exhaust noise; and some mufflers can even separate and remove the pollutants from exhaust gas
Vacuum generator and suction disc		Vacuum generator produces vacuum by the negative pressure formed due to high speed movement of compressed air; the vacuum suction disc grips workpiece by its internal negative pressure, and it is widely used for handling thin sheets and fragile objects, etc.

Task implementation

1 Draw the graphic symbols of pneumatic components on equipment.

IV Pneumatic Logic Components

Pneumatic logic components complete certain logic functions in the system. Under the effect of input signals, the output signal of logic component has only "0" or "1" state (indicating "On" or "Off", "Yes" or "No", and alike); and the logic component is a switching component (or digital component). It uses compressed air as working medium, and realizes certain logic function by changing the output state of air flow through the actions of internal movable parts (e. g., membrane, valve element) subject to the controlling pressure signals.

There are many pneumatic logic components, and they can be generally classified as follows: high-pressure type (0.2-0.8 MPa), low-pressure type (0.05-0.2 MPa) and micro-pressure type (0.005-0.05 MPa), by working pressure; "YES gate", "NOT gate", "OR gate", "AND gate" and "Bistable" components, by logic function; cut-off type, membrane type and slide valve type, etc., by structure. Table 6-3-2 lists the graphic symbols and functions of some commonly used logic components.

Table 6-3-2 Graphic Symbols and Functions of Commonly Used Logic Components

Type	Symbol	Function
YES gate	$a \longrightarrow s$	Same state is always maintained between input signal and output signal of the component, i. e., there is no output if there is no input, and output exists only there is input
NOT gate	$a \longrightarrow s$	Opposite state is always maintained between input signal and output signal of the component, i. e., there is no output if there is input, and vise versa
OR gate	$a, b \longrightarrow s$	Provided that there are two input ports and one output port, the component always has output if one or two ports have input. The two input ports are not connected to each other at all times
AND gate	$a, b \longrightarrow s$	Provided that there are two input ports and one output port, the output exists only when both input ports have input
OR NOT	$a, b \longrightarrow s$	Basic double input or NOT component having two input ports, where the component has output only when both input ports have no input signal
Inhibit gate	$a, b \longrightarrow s$	Output of signal b is inhibited as long as signal a exists; and b output exists only when a does not exist
Bistable	$a \longrightarrow s_1$, $b \longrightarrow s_2$	When input signal a exists, s_1 has output, and s_2 is connected to exhaust vent. When the a signal disappears, component maintains the output state of s_1. Similarly, when input signal b exists, s_2 has output, and s_1 is connected to exhaust vent. When the b signal disappears, component maintains the output state of s_2. When two input exists at the same time, the component state depends on the state corresponding to the signal whichever is input earlier

same time, A is connected to the end whichever has higher pressure, and the other end is closed automatically. The graphic symbol of shuttle valve (or gate) is shown in the Figure. Shuttle valve is widely used such as in parallel circuits under manual and automatic control. Figure 6-3-5(c) shows the outline of shuttle valve (or gate).

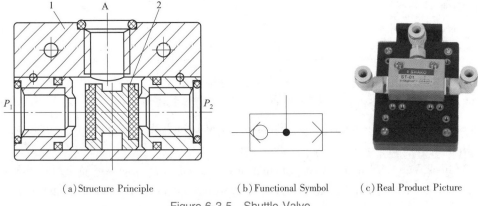

(a) Structure Principle   (b) Functional Symbol   (c) Real Product Picture

Figure 6-3-5   Shuttle Valve

(3) Quick Exhaust Valve. The structure principle of quick exhaust valve is as shown in Figure 6-3-6(a); when compressed air enters air inlet port P, the membrane 1 deforms downward, thereby opening the channel between ports P and A and closing the exhaust vent O. If there is no compressed air entering the air inlet port P, the membrane restores upward under the effect of the differential pressure between ports A and P, thereby closing the port P and discharging air quickly via ports A and O. The graphic symbol of quick exhaust valve is shown in the Figure. Figure 6-3-6(c) shows the outline of quick exhaust valve.

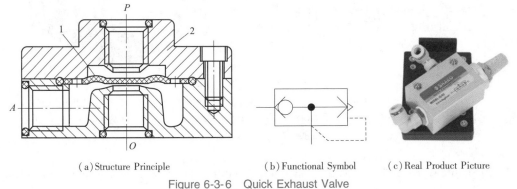

(a) Structure Principle   (b) Functional Symbol   (c) Real Product Picture

Figure 6-3-6   Quick Exhaust Valve

2   Reversing Type Directional Control Valve

Reversing type directional control valve changes the air flow direction by changing the air flow channel, so as to change the movement direction of actuating component. The reversing principle is similar to that of the same type of hydraulic reversing valve, so it is not repeated here.

Table 6-3-1 Classification of Directional Control Valves

Classification	Type
By air flow direction in valve	Check valve, reversing valve
By structure of valve element	Check valve, slide valve
By seal form of valve	Hard seal, soft seal
By working positions and ways of valve	Two-position three-way, two-position five-way, three-position five-way, etc.
By control mode of valve	Pneumatic control, electromagnetic control, mechanical control, manual control

Some typical directional control valves are described as below.

1 One-way Directional Control Valve

One-way directional control valve allows air flow in only one direction. It includes check valve, shuttle valve and quick exhaust valve, etc.

(1) Check Valve. The structure principle of check valve is as shown in Figure 6-3-4(a); when air flow enters via port P, the pressure force overcomes the spring force and the friction force between valve element and valve body to move the valve element leftward, thus the valve port is opened and the air flow passes in normal direction. To ensure the stable air flow, a certain differential pressure shall be maintained between chamber P and chamber A, so as to keep the valve element in open state. When air flow enters chamber A in reversed direction, the valve port is closed, and air flow cannot pass in reversed direction. The graphic symbol of check valve is shown in the Figure 6-3-4(b). Figure 6-3-4(c) shows the outline of check valve.

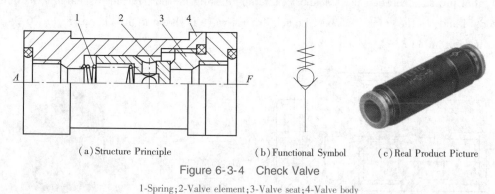

(a) Structure Principle (b) Functional Symbol (c) Real Product Picture

Figure 6-3-4 Check Valve

1-Spring; 2-Valve element; 3-Valve seat; 4-Valve body

(2) Shuttle Valve. The structure principle of shuttle valve is as shown in Figure 6-3-5(a); when two input ports P_1 and P_2 are both required to be connected to output port A, while ports P_1 and P_2 are not allowed to be connected to each other, the shuttle valve (or gate) can be adopted. When air flow enters via P_1, the valve element moves rightward, thus P_1 is connected to A, and air flow comes out through A. At the same time, the valve element closes the P_2 channel. On the contrary, when P_2 is connected to A, the P_1 channel is closed. If air enters via P_1 and P_2 at the

controlled, the switching time of reversing valve and the delivery speed of pneumatic signal shall be controlled sometimes, and all these have to be achieved by regulating the flow rate of compressed air. Flow control valve is the component that controls flow rate by changing the through-flow cross section of valve. Flow control valves comprise throttle valve, one-way throttle valve, exhaust throttle valve, and quick exhaust valve, etc.

Refer to the throttle valve and one-way throttle valve in the hydraulic part for introduction of throttle valve; this section will not repeat the introduction of the same and will simply introduce the quick exhaust valve Figure 6-3-3 (a) shows the working principle of quick exhaust valve. Compressed air enters via the air intake P and pushes the sealed piston upward quickly to open valve port 2 and close exhaust port O at the same time, thus the air intake P is connected with the service port A. When there is no compressed air entering via port P, the sealed piston drops quickly under the differential pressure between port A and port P, thus port P is closed, and port A discharges air quickly via port O.

Quick exhaust valve is usually installed between reversing valve and air cylinder. The application of quick exhaust valve in circuit is described. It enables the cylinder exhaust to be quickly discharged without passing through the reversing valve, thereby accelerating the reciprocating movement speed of air cylinder and shortening the work cycle.

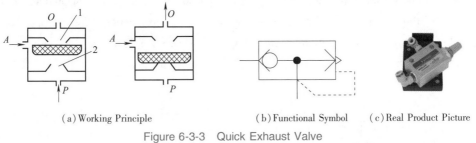

(a) Working Principle (b) Functional Symbol (c) Real Product Picture

Figure 6-3-3 Quick Exhaust Valve

1, 2-Valve ports

III Directional Control Valve

Directional control valve is the pneumatic component that controls the start, stop and movement direction of actuating component by changing the flow direction and on/off of compressed air flow in pneumatic transmission system.

Directional control valves can be classified into several types according to their function, control mode, structure, air flow direction in valve, and seal form, see Table 6-3-1.

valve is in cut-off state, and there is no air flow output.

The pressure regulating range of QTY direct-acting relief valve is 0.05-0.63 MPa. To limit the pressure loss caused by the air flow passing through the relief valve, air flow rate passing through the channel in valve is defined as 15-25 m/s.

The relief valve shall be installed as per the air flow direction and the arrow direction indicated on the valve, in accordance with the installation sequence of water-separating gas filter-relief valve atomized lubricator. Pressure shall be regulated from low to high, until the specified pressure is achieved. When the valve is not used, the handle shall be released, in order to prevent the membrane from deformation under pressure.

2 Overflow Valve

Overflow valve releases a portion of compressed air from the exhaust vent when the system pressure exceeds the set value, and keeps the system pressure basically stable during the relief process, so as to realize the overload protection (the valve is also referred to as safety valve). Overflow valves comprise direct-acting type and pilot type. They can be classified by structure into piston type, membrane type and ball type, etc.

The structure principle of direct-acting overflow valve is as shown in Figure 6-3-2(a); when the input pressure exceeds the set value, valve element 3 overcomes the upper spring force under the pressure force of lower chamber and then rises up, so the valve port is opened to discharge a portion of air, thus the pressure drops, achieving the overload protection. The set pressure can be changed by adjusting the pre-tightening force of spring. Figure 6-3-2(b) shows its graphic symbol. Figure 6-3-2(c) shows the outline of direct-acting overflow valve.

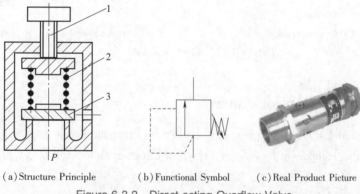

(a) Structure Principle   (b) Functional Symbol   (c) Real Product Picture

Figure 6-3-2   Direct-acting Overflow Valve

1-Adjustment lever; 2-Spring; 3-Valve element

II Flow Control Valve

In a pneumatic transmission system, the movement speed of air cylinder needs to be depressurized through valve port throttle. A portion of the output air flow enter the membrane chamber via damping hole 7 and generates an upward thrust force under the membrane 5; the thrust force always attempts to reduce the valve opening, in order to reduce its output pressure. When the thrust force acting on the membrane is balanced with the spring force, the output pressure of relief valve is maintained constant.

When the input pressure changes, e. g., input pressure increases abruptly, the output pressure increases accordingly, the air thrust force acting on the membrane 5 also increases, thereby compromising the previous force balance and moving the membrane 5 upward; a small amount of air is discharged through overflow port 4 and exhaust vent 11. While the membrane is moving upward, the output pressure drops under the effect of reset spring 10, until a new balance is achieved. After the new balance is achieved, the output pressure basically returns to the former value. On the contrary, when the output pressure drops suddenly, the membrane moves downward, the air intake opening increases, the throttle effect reduces, and the output pressure basically returns to the former value.

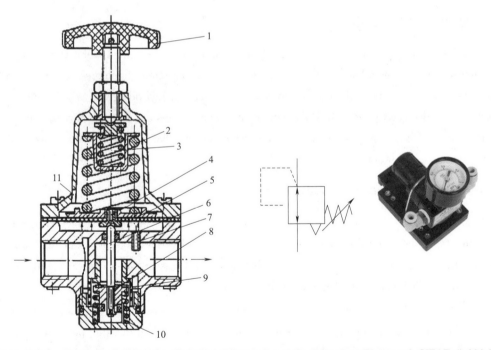

Figure 6-3-1 Structure Schematic, Functional Symbol and Real Product Picture of *QTY* Relief Valve

1-Handle; 2,3-Pressure regulating springs; 4-Overflow port; 5-Membrane; 6-Valve rod;
7-Damping hole; 8-Valve element; 9-Valve seat; 10-Reset spring; 11-exhaust vent

Adjust the handle 1 to restore the free state of springs 2 and 3, the output pressure drops to zero, the valve element 8 closes the air intake under the effect of reset spring 10, therefore the relief

Task implementation

1 Draw the graphic symbols of the pneumatic actuating component on equipment.
2 Skillfully install pneumatic actuating component, with reasonable working table top layout.
3 Discuss by groups, summarize the application scenarios of pneumatic actuating component.

Task 3   Knowledge about Pneumatic Control Components

Task description

Understand the basic working principles and symbols of commonly seen pneumatic control components, be acquainted with the types of pneumatic control valves, and be familiar with the applications of pneumatic control valves in different scenarios.

Relevant knowledge

In a pneumatic transmission system, the pneumatic control components refer to various control valves that control and regulate the pressure, flow rate and direction of compressed air, in order to ensure that the pneumatic actuating components (e. g., air cylinder, air motor, etc.) work properly as per designed programs. Pneumatic control components can be classified by function into pressure control valve, flow control valve, directional control valve and logic components capable of realizing certain logic functions.

I   Pressure Control Valve

Pressure control valve controls the pressure of compressed air, and controls the action sequence of actuating components by the air pressure. Pressure control valve works by balancing the force of compressed air acted on valve element and the spring force. Pressure control valves mainly comprise relief valve, overflow valve and sequence valve.

1 Relief Valve (Pressure Regulating Valve)

Relief valve regulates the outlet pressure to a set valve lower than the inlet pressure, and keeps the output pressure stable (also referred to as pressure regulating valve). Relief valves comprise direct-acting type and pilot type.

Figure 6-3-1 is the structure diagram of QTY direct-acting relief valve. Its working principle is that: when the valve is in operation, adjust the handle 1, pressure regulating springs 2 and 3 and the membrane 5 to move the valve element 8 downward through valve rod 6, thereby the air inlet port is opened, pressure air flow enters from the left end and goes out from the right end after it is

enters the bottom of vane through the slots in the seal caps at both ends, so as to push the vane out and make it stick close to the inner wall of stator; a large portion of the air enters corresponding confined spaces to act on the two vanes; a differential torque is generated because the lengths of the two vanes are not equal, thereby the vanes and the rotors rotate counterclockwise. The acted air is discharged via the holes C and B in the stator. If the input direction of compressed air is changed (namely, input the compressed air via hole B, and discharge via holes A and C), the direction of rotor can be changed.

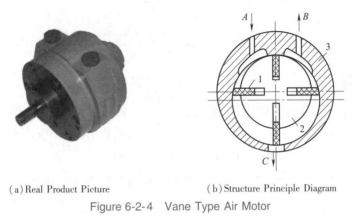

(a) Real Product Picture　　　　　(b) Structure Principle Diagram

Figure 6-2-4　Vane Type Air Motor

Table 6-2-2 lists the characteristics and application ranges of various air motors, for reference while making choices and taking actions.

Table 6-2-2　Characteristics and Application Ranges of Various Air Motors

Type	Torque	Speed	Power	Air consumption per kilowatt/ ($m^3 \cdot min^{-1}$)	Characteristic and application range
Vane type	Low torque	High speed	From a few tenths of kilowatt to 1.3 kW	Small size: 1.8-2.3　Large size: 1.0-1.4	Easy manufacturing, compact structure, but small starting torque at low speed, poor low-speed performance; suitable for machines requiring medium or low power, e.g., handheld tool, combination tool conveyor belt, lift, pump, tractor, etc.
Piston type	Medium high torque	Low- or medium-speed	From a few tenths of kilowatt to 1.7 kW	Small size: 1.9-2.3　Large size: 1.0-1.4	Large power output and good torque characteristics at low speed. Accurate startup; better start and stop characteristics than vane type; suitable for machines having large load and high low-speed torque requirements, e.g., handheld tool, crane, winch, capstan, tube drawing machine, etc.
Membrane type	High torque	Low speed	Less than 1 kw	1.2, u1.4	Suitable for machines requiring accurate control, extremely high starting torque and low speed

Table 6-2-1 Functional Symbols of Commonly Seen Air Cylinders

Single-acting cylinder	Double-acting cylinder		
	Ordinary cylinder	Cushion cylinder	
Spring pressed out	Single piston rod	Non-adjustable one-way	Adjustable one-way
Spring pressed in	Double piston rods	Non-adjustable two-way	Adjustable two-way

II Air Motor

Air motor is also a pneumatic actuating component. Its function is equivalent to an electric motor or a hydraulic motor, namely, the output torque drives the mechanism for rotary motion.

1 Classification and Characteristics of Air Motors

Air motors can be classified by structure into vane type air motor, piston type air motor, and gear type air motor, etc. Piston type air motor and vane type air motor are most commonly seen. Vane type air motor is simple to manufacture and has compact structure, but it has small torque at low speed and poor low-speed performance. It is suitable for medium- and low-power machines, and now popularly used in mining and pneumatic tools. Piston type air motor hasgreat output power at low speed; it has good low-speed performance, and is suitable for machines having large load and low-speed torque requirements, e. g. , crane, winch, capstan, tube drawing machine, etc.

Compared with hydraulic motor, the air motor has the following characteristics:

(1) Safe operation. Suitable for operation in flammable and explosive scenarios, free from influence of high temperature and vibration;

(2) Be able to work under full load for a long period, with a small temperature rise;

(3) Support stepless speed regulation. Revolving speed and power of the motor can be adjusted by controlling the air inlet rate. Rated revolving speed can be from dozens rounds per minute to hundreds of thousands of rounds per minute;

(4) High starting torque. Be able to directly drive the load;

(5) Simple structure, convenient manipulation, easy maintenance, low cost;

(6) Relatively small output power, where the maximum output power is only 20 kW;

(7) High air consumption, low efficiency, large noise.

2 Working Principle of Air Motor

Figure 6-2-4(a) shows the outline of vane type air motor. Its working principle is as shown in Figure 6-2-4(b), where the compressed air is input via hole A; a small portion of the input air

The membrane of a membrane type air cylinder can be made into a disc type membrane or a flat membrane. The membrane is made from textile-included rubber, steel sheet or phosphor bronze sheet. Textile-included rubber is the commonly used material, where the thickness of rubber is 5-6 mm, or 1-3 mm sometimes. Metallic membrane is used only in the membrane type air cylinder with small stroke.

Compared with a piston cylinder, the membrane type air cylinder has advantages such as simple structure, compactness, easy manufacturing, low cost, convenient maintenance and repair, long service life, small leakage, and high efficiency. However, the membrane has limited deflection, so its stroke is short (generally not more than 40-50 mm); and the output force on piston rod of the cylinder decreases when the stroke increases.

3 Rotary Air Cylinder

Figure 6-2-3 shows the structure principle of a rotary air cylinder. Rotary air cylinder consists of air guide header, cylinder body, piston and piston rod, etc. The cylinder body as well as the cylinder head and the air guide header of such air cylinder can be driven to rotate; the piston and the piston rod can only do reciprocating linear movement; the air guide header is immovable and connected with external pipe.

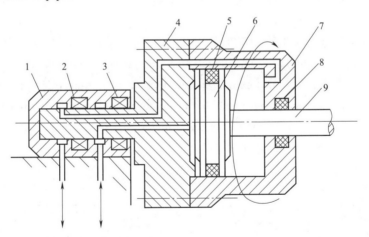

Figure 6-2-3 Rotary Air Cylinder

1-Air guide header; 2,3-Bearings; 4-Cylinder head and air guide header element;
5,8-Seals; 6-Piston; 7-Cylinder body; 9-Piston rod

In the pneumatic system diagram, the air cylinder is indicated by corresponding functional symbol; Table 6-2-1 shows the functional symbols of commonly seen air cylinders:

acting mode; piston type, plunger type, vane type, and membrane type, by structure; ordinary air cylinder and special air cylinder (e.g., impusible type, rotary type and pneumatic-hydraulic damping type), by function.

1 Single-acting cylinder

Figure 6-2-1(a) shows the outline of single-acting air cylinder. Its structure principle is as shown in Figure 6-2-1(b), where the compressed air is sucked in only from one end of air cylinder and pushes the piston (or plunger) to move, and the piston (or plunger) returns by other external force, e.g., spring force, gravity, etc. Single-acting air cylinder is mostly used for short stroke and scenario without high requirements on piston rod push force and movement speed.

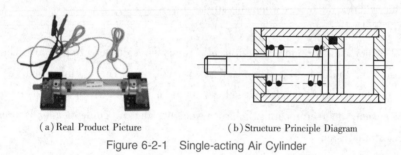

(a) Real Product Picture   (b) Structure Principle Diagram

Figure 6-2-1 Single-acting Air Cylinder

2 Membrane Type Air Cylinder

Membrane type air cylinder is an air cylinder that pushes the piston rod by the membrane driven by compressed air, so as to realize reciprocating linear movement. It consists of the main parts like cylinder body, membrane, membrane disc and piston rod, etc. Its function is similar to that of piston cylinder. The membrane type air cylinder can be classified into single-acting and double-acting types, as shown in Figure 6-2-2.

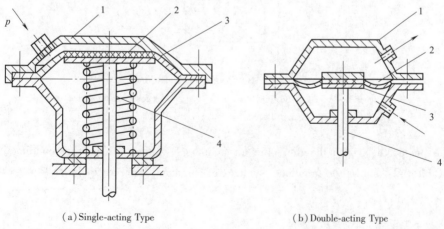

(a) Single-acting Type   (b) Double-acting Type

Figure 6-2-2 Schematic Diagram of Membrane Type Air Cylinder Structure

1-Cylinder body; 2-Membrane; 3-Membrane disc; 4-Piston rod

lubricant once every 36 hours of operation.

(7) Remove the air valve for cleaning once every 500 service hours (or half a year).

(8) Clean the components of machine once every year.

(9) Inspect all protective covers, warning signs and safety devices regularly.

(10) Check the pressure relief device, shutdown protection device and pressure gauge of air compressor regularly (once half a year), and check the sensitivity of safety valve, ensuring that the air compressor is under normal operation conditions.

(11) Regularly check the parts and components exposed to high temperature, e. g., valve, cylinder head, exhaust pipe, clean the oil contamination and carbon deposition on inner wall. During operation, do not touch the high-temperature parts.

Task implementation

1 Simply describe the relation and difference between pneumatic and hydraulic components, and draw the symbol of air supply device.

2 Understand the composition of pneumatic component, and know the constituting parts of each composition.

Task 2   Knowledge about Pneumatic Actuating Components

Task description

Understand the basic working principles and symbols of commonly seen pneumatic actuating components, be acquainted with the types of pneumatic actuating components, and be familiar with the applications of pneumatic actuating components in different scenarios.

Relevant knowledge

Pneumatic actuating component is the device which converts the pressure energy of compressed air into mechanical energy. It includes air cylinder and air motor; the principles of air cylinder and air motor are similar to those of hydraulic cylinder and hydraulic motor. Air cylinder is used for linear reciprocating or swinging movement, and air motor is used for continuous rotating movement.

I Air Cylinder

Air cylinder is one actuating component of pneumatic system. Except some special air cylinders, the types and structures of ordinary air cylinders are basically the same as hydraulic cylinders. Air cylinders are classified as follows: single-acting and double-acting types, by the

When the air supply pressure is greater than 0.1 MPa, the atomized lubricator allows refilling oil without turning off the air circuit. The oil feed quantity changes with the air flow rate. The oil cup and the observation cap are made from transparent material, for easier observation. Atomized lubricator shall be of good leakproofness, pressure resistance and oil drip quantity regulating performance. In use, the parameters like atomizing flow rate shall be reasonably adjusted with reference to relevant standard, in order to achieve the best lubricating effect.

In general, the air of pneumatic system is discharged into atmosphere directly, and the oil content in air is harmful to human; some special industries do not allow the compressed air to contain lubricant. Because of the technological advancement and the application of some new techniques, now some pneumatic components do not need lubricant in the compressed air, thus the air supply regulator consists of only the filter and the relief valve, which are referred to as FR Unit.

2 Symbol of Air Supply Regulator

Figure 6-1-5 shows the real FR Unit and FRL Unit and their symbols; in general, they appear as simplified symbols on system diagrams.

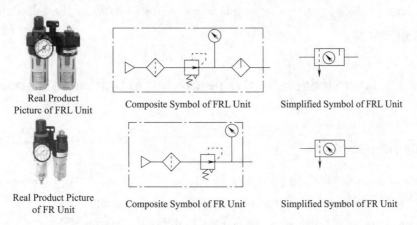

Figure 6-1-5 Real FR Unit and FRL Unit and Symbols Thereof

III Daily Maintenance and Service Precautions of Air Compressor

(1) Keep the machine clean.

(2) Open the drain valve of air storage tank every day to discharge oil water. In humid places, open the valve once every four hours.

(3) Check the lubricant level every day, ensuring the lubrication of air compressor.

(4) Clean or replace the air filter once every 15 days (filter element is consumable).

(5) Check the fixation of screws from time to time.

(6) Replace the lubricant after 50 hours since initial operation, or after one week. The replace the lubricant once every 300 hours (once every 150 hours in severe service environment), and refill

Continued

Constituting part	Functional symbol	Function and effect
Air storage tank		Store the compressed air and eliminate the pressure fluctuation, ensuring the continuous and stable air supply.
Filter		Further remove oil contamination, water and dust from compressed air, in order to promote the working efficiency of downstream drier and extend the service life of fine filter.
Dryer		Further remove the water, oil and dust from compressed air.
Fine filter		Remove the oil contamination, water and dust from compressed air again.

II Air Supply Regulator

In actual operation, the compressed air output from the air compression station cannot meet the requirements of pneumatic component on air supply quality. To fulfill the air supply quality requirement, the air supply regulator is usually installed before the pneumatic system.

1 Composition of Air Supply Regulator

Composition of air supply regulator is as shown in Figure 6-1-4; it consists of three parts, including filter, relief valve and atomized lubricator that are referred to as FRL Unit.

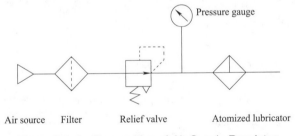

Figure 6-1-4 Composition of Air Supply Regulator

The air filter removes liquid water drops, oil drops, solid dust particles and other impurities from the compressed air. In general, the air filter consists of an enclosure and a filter element. According to the materials of filter element, the air filter can be classified into paper, textile, ceramic, foam plastic and metallic types. Commonly used air filters are paper and metallic types;

The relief valve regulates the air inlet pressure to the pressure required by the system;

The atomized lubricator atomizes the lubricant and spray it into the air flow inside compressed air pipeline, thereby the lubricant enters the systems along with air and lubricates the surfaces of relative movement parts. Atomized lubricator can be classified into oil spray and micro mist types.

climatic conditions. When a pneumatic system is working, the contents of moisture and solid particulate impurities and alike in compressed air are important factors to the normal operation of system. Compressing and purifying air to provide each piece of equipment with clean and dry compressed air is referred to as the compressed air station or air supply; and Figure 6-1-2 shows the schematic of the operation of air supply device.

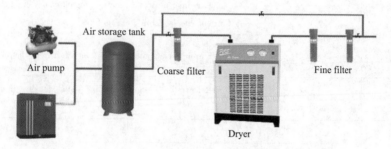

Figure 6-1-2 Schematic of Operation of Air Supply Device

When processing air, the air shall be compressed and treated by a series of filtration and drying operations; Figure 6-1-3 shows the detailed workflow of compressed air station; and Figure 6-1-1 shows the constituting parts of air supply device.

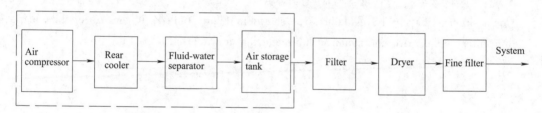

Figure 6-1-3 Workflow of Air Supply Device

Table 6-1-1 Constituting Parts of Air Supply Device

Constituting part	Functional symbol		Function and effect
Air pump			Compress air into compressed air.
Cooler			Cool the compressed air at air compressor outlet to below 40 ℃, to condense most of the vapor and bad oil mist in air as liquid water and oil drops.
Fluid-water separator	Manual		Separate the impurities like water and oil drops dissolved out by aftercooler from the compressed air.
	Auto		

Part 6

Knowledge about Pneumatic Components

Task 1 Air Supply Device

Task description

Air supply device is very similar to hydraulic device, but there are some differences. The task is to understand the basic working principles and symbols of commonly seen air supply devices, be acquainted with the types of air supply devices, and be familiar with the applications of commonly seen air supply devices in different scenarios through the study.

Relevant knowledge

In a pneumatic transmission system, air supply device provides the pneumatic system with compressed air of certain quality, and it is an important part of the pneumatic transmission system. As shown in Figure 6-1-1, the compressed air produced by the air compressor is supplied to the control component and the actuating component after a series of operations including cooling, purification, depressurization and voltage stabilization; and noise is generated when discharging the used compressed air into atmosphere, so corresponding measures shall be taken to reduce the noise, thereby improving the labor conditions and the environmental quality.

Figure 6-1-1 Air compressor

I Air Supply Device

Nature air is a mixture which consists of mainly nitrogen, oxygen, vapor as well as some other trace gases and impurities. The air composition is different under different environmental and

e. g. , petroleum refining, gas processing, chemical fertilizer production, etc.

4 Robot

For examples, assembling robot, painting robot, transfer robot, wall climbing robot and welding robot, etc.

5 Light industry and food packaging

Include various semi-automatic or full-automatic packaging lines, e. g. , wines, oils, gas cans, and food packaging, etc.

6 Others

For examples, vehicle brake, door opening/closing device, particulate material screening, automatic control device of torpedo and missile, etc.

Task implementation

1 Query the "Pneumatic Schematic of Car Door Opening/Closing Device" through network.

2 Grouping and discussion: Group in 3-5 persons, discuss why pneumatic is most widely used in light industry, discuss its principles, and debrief group by group.

Thinking and Exercise

1 What is pneumatic transmission? What is the basic principle of pneumatic transmission?

2 What are the constituting parts of a pneumatic transmission system? What is main content of each constituting part?

3 Simple description of the characteristics and applications of pneumatic transmission.

signal by using compressed air as working medium. It is the technology that control and drives various machines and equipment to realize the mechanization and automation of production process. Because the compressed air as working medium is featured in fire prevention, explosion prevention, electromagnetic interference prevention, vibration prevention, shock prevention, radiation prevention, pollution prevention, simple structure, reliable operation, etc., the pneumatic technology, supplemented by hydraulic, mechanical, electrical and electronic technologies, has evolved into an important approach to realize the automated production process, and has been widely used in many sectors such as mechanical industry, metallurgical industry, light textile and food industry, chemical engineering, transportation, space and aviation, national defense, etc.

Pneumatic transmission is generally used as follows:

1　Machine building industry

Include clamping and handling workpieces on machining production line, modeling, tamping, phase bonding and alike on foundry production line. Automatic handling and fixing, automatic welding and the like of car body components on automated vehicle production line in automobile manufacturing industry. Figure 5-3-1 shows the pneumatic transmission system of a pneumatic drilling machine.

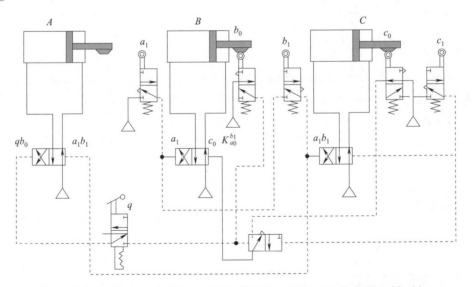

Figure 5-3-1　Pneumatic Transmission System of Pneumatic Drilling Machine

2　Electronic IC and electrical appliance industry

For example, silicon wafer handling, plug-in mounting and welding of components, assembling of home appliance, etc.

3　Petrochemical engineering

Most automated processes transmitting medium through pipelines use pneumatic controls,

process; however, for general mechanical equipment, the transmission speed of pneumatic signal can fulfill the operational requirements.

(4) The noise the large; in particular, muffler shall be provided for supersonic exhaust. Performance comparison between pneumatic transmission and other transmissions see Table 5-2-1.

Table 5-2-1 Performance Comparison Between Pneumatic Transmission and Other Transmissions

Type		Operating force	Action speed	Environmental requirements	Construction	Load change influence	Operating distance	Stepless speed regulation	Service life	Maintenance	Price
Pneumatic transmission		Moderate	Fast	Good suitability	Simple	Large	Medium range	Preferable	Long	General	Cheap
Hydraulic transmission		Maximum	Slow	Free from vibration impact	Complex	A few	Short range	Good	General	High requirement	A little expensive
Electric transmission	Electrical	Moderate	Fast	High requirement	A little complex	Barely	Long range	Good	Short	Relatively high requirement	A little expensive
	Electronic	Minimum	Fastest	Very high requirement	Most complex	No	Long range	Good	Short	Higher requirement	Most expensive
Mechanical transmission		Relatively large	General	General	General	No	Short range	Difficult	General	Simple	General

Task implementation

1 Query the advantages and shortages of commonly used pneumatic transmission through network, and give examples of the advantages and shortages in real life.

2 Grouping and discussion: Discussion the long range pneumatic transmission by groups of 3-5 persons, e.g., the tool changing device of numerically controlled lathe; at report the findings by each group.

Task 3 Applications of Pneumatic Transmission System

Task description

Pneumatic transmission is featured in cleanliness and speediness, so what are the main applications of pneumatic transmission system, e.g., vehicle door opening/closing device? What are the main characteristics and applications of the transmission?

Relevant knowledge

Pneumatic transmission, also referred to as pneumatic for short, is to deliver power and control

(1) The working medium is air that is inexhaustible, so energy resource can be saved as compared with hydraulic fluids. Air is uneasy to block the flow channel, and can be discharged into atmosphere at any time after use, without resulting in environmental pollution.

(2) Temperature has small influence on characteristics of air. Air can work reliably under high temperature, without leading to combustion or explosion. When temperature changes, there is little influence on air viscosity, thus the transmission performance is not affected.

(3) Air viscosity is very small (about 1/10,000 of that of hydraulic fluid), so the flow resistance is small, and the pressure loss is small when flowing in pipeline; therefore, air is convenient for centralized supply and long-distance supply.

(4) As compared to hydraulic transmission, pneumatic transmission acts and responds quickly; in general, the working pressure and speed can be achieved just in 0.02-0.3 s. The flow speed of hydraulic fluid in pipeline is generally 1-5 m/s, while the air flow speed is at least more than 10 m/s or even up to the speed of sound; the exhaust speed can further be up to the supersonic speed.

(5) Air pressure has strong self-sustaining capability; even if the compressor has stopped and the air valve is closed, a stable pressure can still be maintained in the device. However, to maintain the pressure in a hydraulic system, the energy pump shall keep working continuously or additional energy accumulator shall be provided; but air maintains the constant pressure of load cylinder by its own expansibility.

(6) Pneumatic components have high reliability and long service life. Electrical components can be operated for a million cycles, but pneumatic components can be operated for 20-40 million cycles.

(7) The work environment suitability is good, especially in severe environments such as flammable, explosive, dusty, strong magnetic, radiative, vibrating and other environments. Pneumatic transmission is superior to hydraulic, electronic and electrical transmissions and controls.

(8) Pneumatic device has simple structure, low cost, convenient maintenance, and automatic overload protection.

II Shortages of Pneumatic Transmission

(1) As air is highly compressible, the pneumatic device has poor action stability; and the change of external load has great influence on working speed.

(2) Because the working pressure is low, the output force of torque of pneumatic device is limited. If the structural dimensions are the same, the output force of pneumatic transmission device is much smaller than that of hydraulic transmission device. The output force of pneumatic transmission device shall not be greater than 10-40 kN.

(3) Signal transmission speed in pneumatic device is slower than optical and electrical controls, so the pneumatic device is not suitable for complex lines having high requirements on signal transmission speed. Meanwhile, it is difficult to realize the remote control of production

(3) Actuating component, referring to an energy converting device that converts the air pressure energy into mechanical energy. It includes the air cylinder for realizing linear reciprocating movement and the pneumatic motor or swing motor for realizing continuous rotating or swinging movement;

(4) Auxiliary component, which is necessary to ensure the purification of compressed air, the lubrication of components, the connection and muffling between components, etc., including filter, atomized lubricator, pipe joint and muffler, etc.

Task implementation

1 Query what transmission systems are using pneumatic transmission through network, and give three examples.

2 Grouping and discussion: Group in 3-5 persons, and discuss the characteristics of pneumatic transmission, and explain the difference from other transmissions. At last, report the findings group by group.

3 Query the development of pneumatic system in China, give objective assessment, and give your own opinions.

Task 2 Characteristics of Pneumatic Transmission System

Task description

What are the advantages and shortages of pneumatic transmission? Look up data to find out the differences between pneumatic transmission and other transmissions.

Relevant knowledge

I Advantages of Pneumatic Transmission

Pneumatic technologies are widely used in industrial sectors in China, such as machinery, electronics, light industry, textile, food, medicine, packaging, metallurgy, petrochemical, aviation, transportation, etc. Pneumatic manipulator, combined machine tool, machining center, automated production line, automated detection and testing devices has sprung up in large numbers, and exhibited great advantages in promoting production efficiency, automation degree, product quality, operational reliability as well as in realizing special processes. This is mainly because that the pneumatic transmission has the following characteristics as compared with conventional mechanical, electrical and hydraulic transmissions.

system can also be used for explosive and flammable scenarios as well as the scenarios subject to impacts.

The pneumatic manipulator shown in Figure 5-1-1 is one of the important devices of automatic production equipment and production line. It acts according to predetermined control programs based on the operational needs of various kinds of automated equipment. Therefore, it is widely used in machining, stamping, forging, casting, assembling, heat treatment and the like for handling workpieces, so as to reduce the labor intensity of workers. It can also be used for achieving the functions such as automatic material fetching, loading, unloading and automatic tool changing. Pneumatic manipulator is a kind of manipulator, which is featured in simple structure, small dead weight, quick action, stability, reliability and energy-saving, etc.

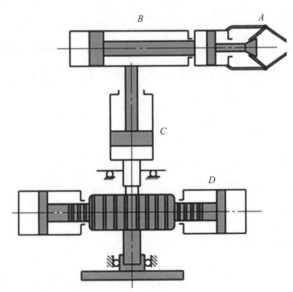

Figure 5-1-1　Pneumatic Manipulator

‖　Composition of Pneumatic Transmission System

Similar to the hydraulic transmission system, the pneumatic transmission system also consists of four parts, including:

(1) Air supply device, which is the device for obtaining compressed air. Its main part is an air compressor that converts the mechanical energy of prime mover into air pressure energy'

(2) Control component, which is used for controlling the pressure, flow rate and flow direction of the compressed air, such that the actuating mechanism can complete predetermined working cycle; these components include various pressure control valves, flow control valves and directional control valves, etc. ;

Part 5
Pneumatic Transmission Fundamentals

Task 1 Working Principle and Composition of Pneumatic Transmission System

Understand the working principle of pneumatic transmission as well as the composition of pneumatic transmission system. What are the main application scenarios of commonly used pneumatic transmissions?

I Working Principle of Pneumatic Transmission System

Pneumatic transmission is a technique which delivers energy and signal by using compressed air as working medium. The working principle of pneumatic transmission that an air compressor converts the mechanical energy output from an electric motor or other prime mover into the air pressure energy, then actuating component converts the pressure energy into mechanical energy in the form of linear or rotating movement under the effect of control component, so as to complete various actions and to do external work. Pneumatic transmission uses compressed air as transmission medium; air can be conveniently obtained and the used gas can also be conveniently discharged, so the series problems of pollution caused by leakage of hydraulic transmission can be avoided. At present, pneumatic technology has evolved quickly from heavy industries such an automobile, mining, steel and machinery to the industries like chemical engineering, light industry, military, etc. Because the medium is air, pneumatic control system is especially suitable for production of food, radio components and medicine. By choosing specific gas medium, the pneumatic control

Thinking and Exercise

1　What is the advantage of unloading the pump after the working parts have stopped moving in hydraulic system?

2　How to adjust the speed of the actuator?

Part 4 Installation and Commissioning of Basic Hydraulic Circuits | 095

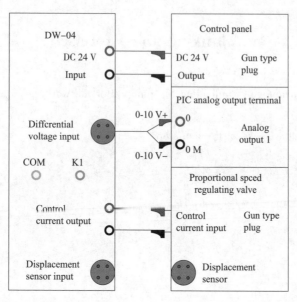

Figure 4-5-4 Siemens System Proportional Speed Regulating Valve Wiring Diagram

Power on the equipment, press the [Start] button on the control panel, and download corresponding proportional speed regulating valve PID control program into corresponding PLC. Turn the vane pump two-position knob switch in pump station electric cabinet to [On] position to start the variable vane pump. Press the *SB2* and the *SB3* button switches to respectively set the hydraulic motor speed to 50 r/min and 80 r/min, and observe the speed change of hydraulic motor; if it is unfavorable, adjust relevant parameters of PID; press the SB1 button to clear the speed. After the practical training, turn the two-position knob switches of vane pump, DC 24 V power supply and AC 220 V power supply to [Off] positions, press the [Stop] button on the control panel, and remove the connecting wires to finish the practical training.

4 Precautions

To download PLC programs, corresponding PLC power supply must be cut off before plugging or unplugging the connecting wires.

Task implementation

1 Grouping and discussion: Group in 3-5 persons, then discuss the installation and commissioning precautions according to the installation and commissioning of the hydraulic circuit of proportional speed regulating valve.

2 Discuss the gains from the hydraulic circuit commissioning by groups.

Turn the DC 24 V power supply two-position knob switch on control panel to [On] position, the "feedback break" indicator on the DW-04 proportional speed regulating valve control module is lit up, and the "manual voltage input" indicator is lit up; rotate the "input setting" potentiometer knob clockwise by certain turns, the hydraulic motor starts to rotate; continuously adjust the potentiometer knob, the motor speed increases; rotate the potentiometer knob counterclockwise, the motor speed reduces continuously. After the practical training, rotate the DW-04 potentiometer knob counterclockwise to zero position, unplug the connecting wires between COM and K1, turn the two-position knob switches of vane pump, DC 24 V power supply and AC 220 V power supply to [Off] positions, and press the [Stop] button on the control panel.

3 Practical Training for PID Control of Proportional Speed Regulating Valve Based on PLC

Check the wire connections according to Figure 4-5-3 Siemens System PLC Wiring Diagram and Figure 4-5-4 Siemens System Proportional Speed Regulating Valve Wiring Diagram, power on the equipment, and turn the DC 24 V power supply and AC 220 V power supply two-position knob switches on the control panel to [On] positions.

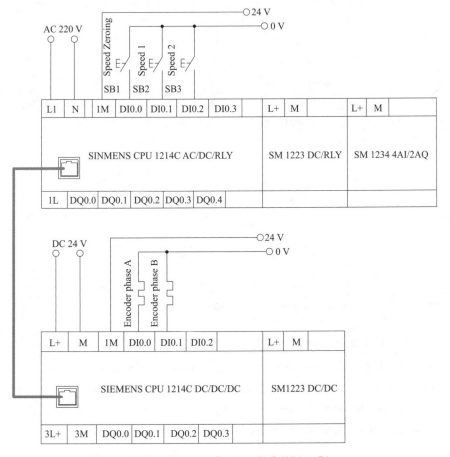

Figure 4-5-3 Siemens System PLC Wiring Diagram

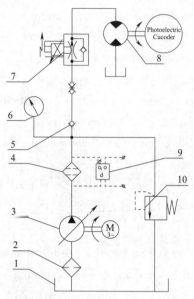

Figure 4-5-1 Control Hydraulic Schematic of Proportional Speed Regulating Valve

1-Oil tank; 2-Oil suction filter; 3-Variable vane pump; 4-High pressure filter; 5-Check valve;
6-Pressure gauge; 7-Proportional speed regulating valve; 8-Hydraulic motor;
9-Differential pressure annunciator; 10-Direct-acting overflow valve

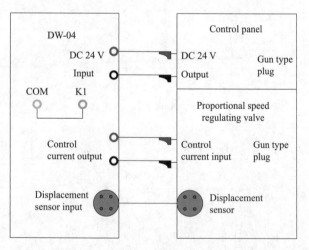

Figure 4-5-2 Manual Control Wiring Diagram of Proportional Speed Regulating Valve

2 Practical Training for Manual Control of Proportional Speed Regulating Valve

Power on the equipment, press the [Start] button on the control panel, turn the vane pump two-position knob switch in pump station electric cabinet to [On] position to start the variable vane pump, adjust the handle of direct-acting overflow valve 10 clockwise, and adjust the system pressure to 4 MPa. After regulated the system pressure, lock up the locking nut.

Task 5 Installation and Commissioning of Proportional Valve Control Circuits

 Task description

Proportional control valve refers to the valve that converts the small-power electric current input into amplified hydraulic power (load pressure and load flow) output by amplifying components in electrohydraulic servo system, in order to control the displacement, speed, acceleration and force of actuating components. What are the commonly used proportional control valves? What are the installation precautions?

 Relevant knowledge

I Composition of Proportional Control Valve

Proportional control valve mainly consists of three parts: electrical-mechanical converter, hydraulic amplifier and feedback mechanism. The electrical-mechanical converter converts the electric signal input into corner or linear displacement output; the corner output device is referred to as a torque motor; and the linear displacement device is referred to as a force motor.

The hydraulic amplifier receives the small-power corner or linear displacement signal input from the electrical-mechanical converter, regulates and distributes the large-power hydraulic fluid, thereby converting and amplifying the control power. The feedback and balance mechanism is that the flow or pressure output from electrohydraulic servo valve obtains the characteristics that are proportional to the input electric signal.

II Connection and Commissioning of Proportional Speed Regulating Valve

1 Connection of Proportional Speed Regulating Valve

Connect the oil path according to Figure 4-5-1 Control Hydraulic Schematic of Proportional Speed Regulating Valve, and turn the pressure regulating handle of direct-acting overflow valve 10 counterclockwise to the maximum opening. Connect wires according to Figure 4-5-2 Manual Control Wiring Diagram of Proportional Speed Regulating Valve, and turn the knob of "input setting" potentiometer counterclockwise to zero position. Turn the manual/auto two-position knob switch in the pump station electric cabinet into [manual] position, and turn the plunger pump and vane pump two-position knob switch to [Off] position.

The real sandwich hydraulic lock used in the example of Figure 4-4-6 is as shown in Figure 4-4-7(a).

(a) Sandwich Hydraulic Lock　　　(b) Y Type Three-position Four-way Solenoid Reversing Valve

Figure 4-4-7　Sandwich Valve Bank

Comparing with the connection and commissioning of the hydraulically controlled locking circuit of the plate valve shown in Figure 4-4-8, it can be found that many pipes can be omitted by using the sandwich hydraulically controlled locking circuit, and it is not easy to make mistakes when connecting the pipes.

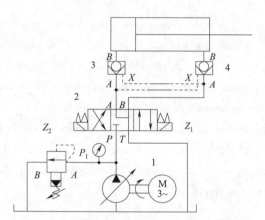

Figure 4-4-8　Schematic Diagram of Hydraulically
Controlled Locking Circuit of Plate Valve

Task implementation

1　As to the hydraulically controlled locking circuit shown in Figure 4-4-8, why the Y type three-position four-way reversing valve is used, and can the self-locking still be achieved if it is changed into O type?

2　Look up materials to see whether the use of sandwich valve bank has any shortage?

3　Are there any other type of valve than sandwich valve, e.g., cartridge valve? Find out the application scenarios of cartridge valve.

(a) MSA Type Sandwich Oil-return Throttle Valve

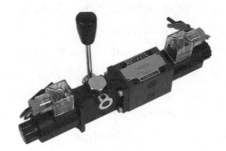
(b) Manual *H* Type Three-position Four-way Solenoid Reversing Valve

Figure 4-4-5 Sandwich Valve Bank

3 Sandwich Locking Circuit

Locking circuit enables a working part to stay in any location, and prevents the part against movement under stress if the work is stopped.

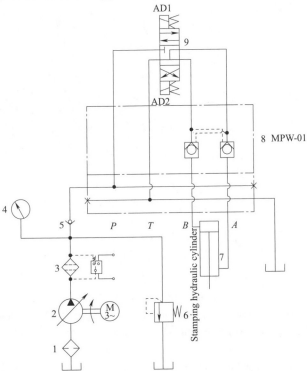

Figure 4-4-6 Hydraulically Controlled Check Valve Locking Circuit

1-Oil suction filter; 2-Variable vane pump; 3-High pressure fluid filter; 4-Shock-proof pressure gauge;
5-Check valve; 6-Direct-acting overflow valve; 7-Stamping hydraulic cylinder;
8-Sandwich hydraulic lock; 9-Three-position four-way reversing valve (Y type)

2 Sandwich Oil-return Throttle Speed Regulating Circuit

Figure 4-4-4 shows the oil-return throttle speed regulating circuit of one-way throttle valve, where the one-way throttle valve 8 is located on the oil return path of actuating mechanism 7 and suitable for the scenario that the working actuating components generate load or the load reduces suddenly. The output pressure of hydraulic pump is the regulated pressure of overflow valve that is not related with the load, thus the efficiency is low. However, it is able to generate backpressure, prevent the occurrence of negative load and prevent onrush.

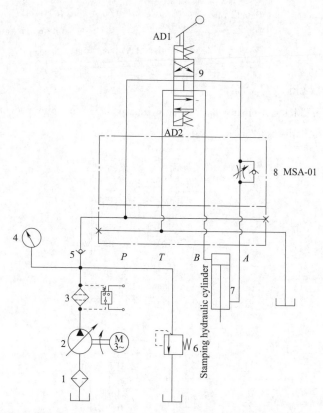

Figure 4-4-4 Sandwich Oil-return Throttle Speed Regulating Circuit

1-Oil suction filter; 2-Variable vane pump; 3-High pressure fluid filter; 4-Shock-proof pressure gauge;
5-Check valve; 6-Direct-acting overflow valve; 7-Stamping hydraulic cylinder;
8-Sandwich throttle valve; 9-Three-position four-way reversing valve (H type)

The real sandwich throttle valve used in the example of Figure 4-4-4 is as shown in Figure 4-4-5(a), where the flow rate of oil return can be regulated by adjusting the knob on the left side. The hydraulic cylinder extending speed can be regulated thereby. In addition, the manual H type three-position four-way solenoid reversing valve as shown in Figure 4-4-5(b) is used for commutation.

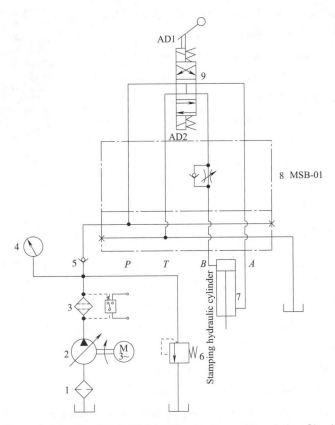

Figure 4-4-2 Sandwich Oil-inlet Throttle Speed Regulating Circuit

1-Oil suction filter; 2-Variable vane pump; 3-High pressure fluid filter; 4-Shock-proof pressure gauge;
5-Check valve; 6-Direct-acting overflow valve; 7-Stamping hydraulic cylinder;
8-Sandwich throttle valve; 9-Three-position four-way reversing valve (H type)

(a) MSB Type Sandwich Oil-inlet Throttle Valve

(b) Manual H Type Three-position Four-way Solenoid Reversing Valve

Figure 4-4-3 Sandwich Valve Bank

completely the same as that of ordinary hydraulic valve. However, the sandwich valves have special structure and connection characteristics, so they form a separate series. Each sandwich valve in a system works not only as the control component, but also as the channel. Sandwich valves of the same diameter, including various pressure valves, flow valves, directional valves, etc., consist of one series. In the same series, the oil port location, screw hole location and connection size of each valve must be identical, so various typical hydraulic systems can be formed just be superposing each valve in specific orders.

Figure 4-4-1 Triple Sandwich Valve Bank

II Assembling of Sandwich Valve

Assembling requirements of sandwich valve are basically the same as the requirements of conventional plate check valve. However, special attentions shall be pay to the match of mounting surface of each valve, in order to prevent leakage. Assembling steps are as below:

(1) According to the design of hydraulic circuit, superpose the sandwich valve and the solenoid reversing valve at first, where the side of sandwich valve with O-ring shall face the base plate.

(2) Before inserting bolts into the sandwich valve, confirm whether the oil port location of sandwich valve is correct, and align the screw mounting holes of sandwich valve.

(3) Insert the special mounting screws of sandwich valve, ensuring that every screw is fastened with specified torque.

Figure 4-4-1 shows the properly assembled triple sandwich bank.

III Commonly Used Sandwich Circuits

1 Sandwich Oil-inlet Throttle Speed Regulating Circuit

Figure 4-4-2 shows the oil-inlet throttle speed regulating circuit of one-way throttle valve, where the one-way throttle valve 8 is located on the oil-inlet path of actuating mechanism 7 and suitable for the hydraulic cylinder operated under positive load. This circuit has low efficiency, large speed regulating range, and is suitable for low load and low speed working conditions. The quantity of oil pipes required in this circuit is greatly reduced, as compared to the oil-inlet throttle speed regulation consisting of front panel type valve.

The real sandwich throttle valve used in the example of Figure 4-4-2 is as shown in Figure 4-4-3(a), where the flow rate of oil inlet can be regulated by adjusting the knob on the right side. The hydraulic cylinder extending speed can be regulated thereby. In addition, the manual H type three-position four-way solenoid reversing valve shown in Figure 4-4-3(b) is used for commutation.

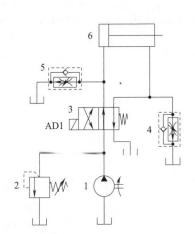
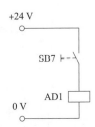

(a) Bidirectional Bypass Throttle Speed Regulating
Circuit of Speed Regulating Valve

(b) Schematic Diagram of
Electrical Control

Figure 4-3-8   Bidirectional Bypass Throttle Speed Regulating Circuit of Speed Regulating Valve
1-Hydraulic pump; 2-Direct-acting overflow valve; 3-Two-position four-way solenoid reversing valve;
4,5-Speed regulating valves; 6-Double-acting hydraulic cylinder with single piston rod

Task 4   Installation and Commissioning of Sandwich Valve Control Circuits

Task description

Understand the advantages and shortages of sandwich valves and their assembling procedures, and what are the commonly used sandwich valve circuits? There are many types of valves used as controls in sandwich valve bank, what is the difference of each type? What else valve banks are there?

Relevant knowledge

I   Introduction to Sandwich Valve and System

The sandwich valve is a novel component developed on the basis of integrated package and manifold. It is featured in compact structure, small size, simple maintenance and repair, etc. A hydraulic system consisting of sandwich valves do not need additional connectors, as the system is formed by superposing the valves that are connected by their own bodies, as shown in Figure 4-4-1. Such system avoids the problems such as leakage, vibration and noise caused by pipeline, joint and flange, and significantly simplifies the hydraulic system. Working principle of sandwich valve is

regulating valve, the circuit is a bypass throttle speed regulating circuit of speed regulating valve.

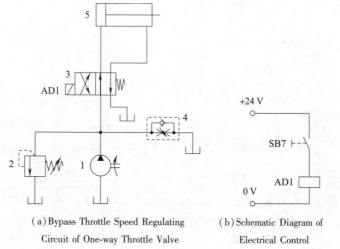

(a) Bypass Throttle Speed Regulating (b) Schematic Diagram of
Circuit of One-way Throttle Valve   Electrical Control

Figure 4-3-7 Bypass Throttle Speed Regulating Circuit of One-way Throttle Valve

1-Hydraulic pump; 2-Direct-acting overflow valve; 3-Two-position four-way solenoid reversing valve;
4-One-way throttle valve; 5-Double-acting hydraulic cylinder with single piston rod

Comparison of the advantages and shortages of the three circuits above are given in Table 4-3-1.

Table 4-3-1 Comparison of Speed Regulating Circuits

Main performance of circuit		Throttle speed regulating circuit			
		Throttle valve		Speed regulating valve	
		Oil inlet and return	Bypass	Oil inlet and return	Bypass
Mechanical characteristics	Speed stability	Relatively poor	Poor	Good	
	Bearing capacity	Preferable	Relatively poor	Good	
Speed regulation range		Relatively large	Small	Relatively large	
Power characteristics	Efficiency	Low	Relatively high	Low	Relatively high
	Heat	Large	Small	Large	Small
Scope of Application		Medium- and low-pressure systems with small power and small load			

Task implementation

1 Figure 4-3-8 shows the bidirectional bypass throttle speed regulating circuit of speed regulating valve, in order to analyze its working principle and characteristics.

2 After analysis, discuss the difference between the bidirectional bypass oil path throttle speed regulating circuit and the one-way bypass throttle speed regulating circuit by groups.

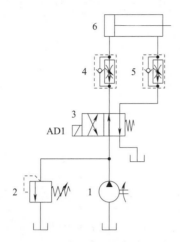
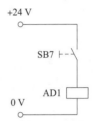

(a) Bidirectional Oil-return Throttle Speed Regulating Circuit of Speed Regulating Valve

(b) Schematic Diagram of Electrical Control

Figure 4-3-5 Bidirectional Oil-return Throttle Speed Regulating Circuit of Speed Regulating Valve

1-Hydraulic pump; 2-Direct-acting overflow valve; 3-Two-position four-way solenoid reversing valve; 4,5-Speed regulating valves; 6-Double-acting hydraulic cylinder with single piston rod

3 Bypass Throttle Speed Regulating Circuit

Figure 4-3-6 shows the load characteristics curve of bypass throttle speed regulating circuit; when the throttle valve opening increases, the piston movement speed reduces; when the throttle valve opening is not changed, the piston movement speed reduces quickly while the load increases, and its speed rigidity is lower than that of the oil-inlet and oil-return throttle speed regulating circuits. When the load is constant, the throttle valve opening is smaller, the speed rigidity is higher, and the tolerable maximum load is greater. The working pressure of hydraulic pump changes along with the load change; there is only throttle loss and no overflow loss in the circuit; therefore, this circuit has higher efficiency because there is no backpressure in the oil return chamber of hydraulic cylinder; thus the movement stability is low, and it cannot bear the negative load.

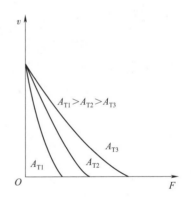

Figure 4-3-6 Load Characteristics Curve of Bypass Throttle Speed Regulating Circuit

Figure 4-3-7(a) shows the bypass throttle speed regulating circuit of one-way throttle valve; in the circuit, excess oil is directly discharged into the oil tank via throttle valve 4; the pressure of hydraulic pump changes with the load; the safety valve is opened only when the oil pressure exceeds the safety pressure; therefore, the efficiency is high. If valve 4 is replaced with one-way speed

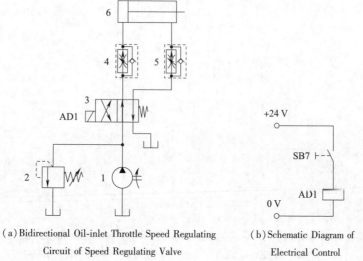

(a) Bidirectional Oil-inlet Throttle Speed Regulating
Circuit of Speed Regulating Valve

(b) Schematic Diagram of
Electrical Control

Figure 4-3-3 Bidirectional Oil-inlet Throttle Speed Regulating Circuit of Speed Regulating Valve

1-Hydraulic pump; 2-Direct-acting overflow valve; 3-Two-position four-way solenoid reversing valve;
4,5-Speed regulating valves; 6-Double-acting hydraulic cylinder with single piston rod

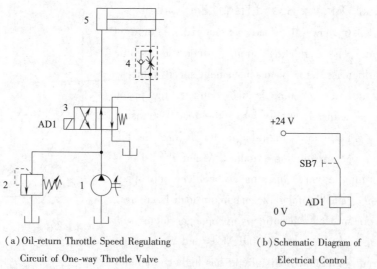

(a) Oil-return Throttle Speed Regulating
Circuit of One-way Throttle Valve

(b) Schematic Diagram of
Electrical Control

Figure 4-3-4 Oil-return Throttle Speed Regulating Circuit of One-way Throttle Valve

1-Hydraulic pump; 2-Direct-acting overflow valve; 3-Two-position four-way solenoid reversing valve;
4-One-way throttle valve; 5-Double-acting hydraulic cylinder with single piston rod

Figure 4-3-5(a) shows the bidirectional oil-return throttle speed regulating circuit of speed regulating valve, in order to analyze its working principle and characteristics.

Figure 4-3-2(a) shows the oil-inlet throttle speed regulating circuit of one-way throttle valve, where the one-way throttle valve 4 is located on the oil-inlet path of actuating mechanism 5 and suitable for the hydraulic cylinder operated under positive load. Excess oil of hydraulic pump is discharged through the overflow valve 2, and the operation is implemented at the set pressure of overflow valve. This circuit has low efficiency, large speed regulating range, and is suitable for low load and low speed working conditions. If valve 4 is replaced with a one-way speed regulating valve, the circuit is an oil-inlet throttle speed regulating circuit, and the speed stability is better than that of throttle valve.

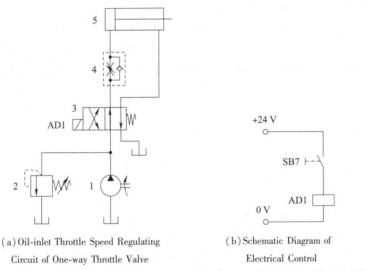

(a) Oil-inlet Throttle Speed Regulating Circuit of One-way Throttle Valve

(b) Schematic Diagram of Electrical Control

Figure 4-3-2　Oil-inlet Throttle Speed Regulating Circuit of One-way Throttle Valve

1-Hydraulic pump; 2-Direct-acting overflow valve; 3-Two-position four-way solenoid reversing valve; 4-One-way throttle valve; 5-Double-acting hydraulic cylinder with single piston rod

Figure 4-3-3 (a) shows the bidirectional oil-inlet throttle speed regulating circuit of speed regulating valve, in order to analyze its working principle and characteristics.

2　Oil-return Throttle Speed Regulating Circuit

Figure 4-3-4(a) shows the oil-return throttle speed regulating circuit of one-way throttle valve, where the one-way throttle valve 4 is located on the oil return path of actuating mechanism 5 and suitable for the scenario that the working actuating components generate load or the load reduces suddenly. The output pressure of hydraulic pump is the regulated pressure of overflow valve that is not related with the load, thus the efficiency is low. However, it is able to generate backpressure, prevent the occurrence of negative load and prevent onrush.

flow rate q is achieved by using a flow valve of a variable pump, and change of the hydraulic motor displacement V_m is achieved by using a variable hydraulic motor. Commonly used speed regulating circuits are described as below.

1 Oil-inlet Throttle Speed Regulating Circuit

Speed load characteristics of hydraulic cylinder, when the hydraulic cylinder works stably, its stress balance equation is

$$p_1 A = F + p_2 A$$

The pump oil supply pressure p_p is regulated to constant by overflow valve, thus the differential pressure between both ends of throttle valve is

$$\Delta p = p_p \; p_1 = p_p - \frac{F}{A}$$

Flow of throttle valve entering the hydraulic cylinder is

$$q_1 = KA_T \Delta p^m = KA_T \left(p_p - \frac{F}{A} \right)^m$$

The speed load characteristics equation is

$$v = \frac{q_1}{A} = \frac{KA_T}{A} \left(p_p - \frac{F}{A} \right)^m$$

The equation above is the speed load characteristics equation of oil-inlet throttle speed regulating circuit. When working under the same load, this circuit has a large range of speed regulation. After regulation, the speed decreases while the load increase, therefore, this speed regulating circuit has low speed load characteristics.

According to the speed load characteristics curve of oil-inlet throttle speed regulation in Figure 4-3-1, when the throttle valve opening A_T is not changed, the piston movement speed v decreases while the load F increase, thus the speed rigidity is low. When the throttle valve opening A_T is certain, the curve of section with small load is gentle and the speed rigidity is high, while the curve of section with large load is steep and the speed rigidity is low. When working under the same load, if the throttle valve opening is small, the piston speed v is low, the curve is gentle and the speed rigidity is high; if the throttle valve opening is large, the piston speed v is high, the curve is steep and the speed rigidity is low.

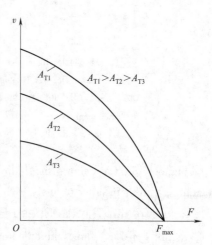

Figure 4-3-1 Speed Load Characteristics Curve of Oil-inlet Throttle Speed Regulation

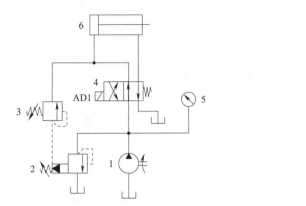

(a) Example Three of Two-stage Pressure Regulating Circuit　　(b) Schematic Diagram of Electrical Control

Figure 4-2-7　Two-stage Pressure Regulating Circuit

1-Hydraulic pump; 2-Pilot type overflow valve; 3-Direct-acting overflow valve;
4-Two-position four-way solenoid reversing valve; 5-Shock-proof pressure gauge;
6-Double-acting hydraulic cylinder with single piston rod

Task 3　Installation and Commissioning of Speed Control Circuits

Task description

Speed control circuits are about the research on speed regulating and changing issues in hydraulic systems; commonly used speed control circuits comprise speed regulating circuit, fast circuit, etc. The purpose is to analyze the working principles and characteristics of bidirectional bypass throttle speed regulating circuit of speed regulating valve.

Relevant knowledge

I　Speed Regulating Circuits

It can be known from the working principle of hydraulic motor that the revolving speed n_m of hydraulic motor depends on the input flow rate and the hydraulic motor displacement V_m, namely, $n_m = q/V_m$, and the movement speed v of hydraulic cylinder depends on the input flow rate and the effective acting area A of hydraulic cylinder, namely, $v = q/A$. According to the relations above, the regulation of the revolving speed n_m of hydraulic motor or the movement speed v of hydraulic cylinder can be achieved by changing the input flow rate q, the hydraulic motor displacement V_m and the effective acting area A of cylinder. Because the effective area A of hydraulic cylinder is a fixed value, the speed can be regulated only by changing the flow rate q, while change of the input

III Depressurizing Circuits

When the pump outputs a high pressure but local circuit or branch requires a low pressure, the depressurizing circuit can be adopted, e. g. , the locating, clamping circuit indexing and hydraulic component control oil paths in machine tool hydraulic system always require pressures lower than the main oil path pressure. Depressurizing circuit is relatively simple, which is generally formed by connecting a relief valve in series with the branch to be depressurized. A stable low pressure of certain branch can be conveniently obtained by using a depressurizing circuit, but the pressure fluid will be subject to pressure loss when it passes through the relief valve.

Figure 4-2-6(a) is a typical depressurizing circuit, which is mainly used in the clamping circuit. The maximum oil supply pressure of hydraulic pump is regulated by the overflow valve 2 according to the demand of main oil path. After the main oil path is connected in parallel with relief valve 3, the clamping cylinder 7 obtains a lower pressure, where the clamping force can be regulated by relief valve 3 and maintained constant. The check valve 4 is used for maintaining pressure in short time, and prevents the clamping force against reducing when the main oil path pressure is lower than the pressure regulated by relief valve.

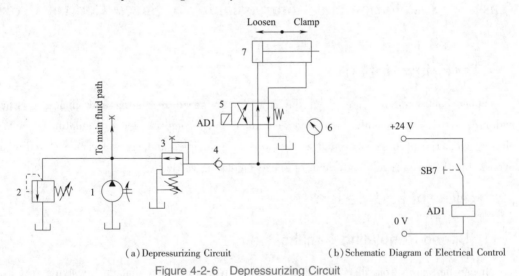

(a) Depressurizing Circuit　　　　　　　(b) Schematic Diagram of Electrical Control

Figure 4-2-6　Depressurizing Circuit

1-Hydraulic pump; 2-Direct-acting relief valve; 3-Direct-acting relief valve; 4-Check valve
5-Two-position four-way solenoid reversing valve; 6-Shock-proof pressure gauge;
7-Double-acting hydraulic cylinder with single piston rod

Task implementation

Figure 4-2-7 is another example of two-stage pressure regulating circuit. Try to analyze the principle of two-stage pressure regulation.

load operation state, thereby reducing the power consumption and reducing the system heat.

1 Unloading Circuit at Remote Control Interface of Pilot Type Overflow Valve

In Figure 4-2-4(a), the remote control interface of pilot type overflow valve 3 is directly connected with two-position three-way solenoid reversing valve 4, in order to form an unloading circuit at remote control interface of pilot type overflow valve. When the reversing valve 4 is in the position shown in the figure, the system pressure of main oil path is regulated by overflow valve 3; when electromagnet AD1 is energized and the solenoid valve 4 changes direction, the remote control interface of overflow valve 3 is connected with the oil tank via valve 4, thereby unloading the system pressure.

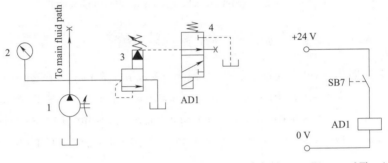

(a) Unloading Circuit at Remote Control Interface of Pilot Type Overflow Valve　　(b) Schematic Diagram of Electrical Control

Figure 4-2-4　Unloading Circuit at Remote Control Interface of Pilot Type Overflow Valve

1-Hydraulic pump; 2-Shock-proof pressure gauge; 3-Pilot type overflow valve;
4-Two-position three-way solenoid reversing valve

2 Reversing Valve Unloading Circuit

In Figure 4-2-5(a), the reversing valve 4 is used for unloading the pump, but there is a great changeover pressure impact; so the circuit is suitable for small-flow systems.

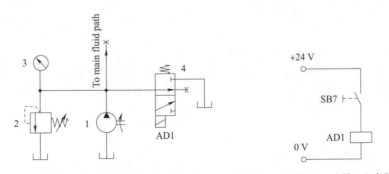

(a) Reversing Valve Unloading Circuit　　(b) Schematic Diagram of Electrical Control

Figure 4-2-5　Reversing Valve Unloading Circuit

1-Hydraulic pump; 2-Direct-acting overflow valve; 3-Shock-proof pressure gauge;
4-Two-position three-way solenoid reversing valve

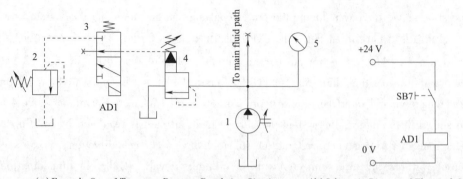

(a) Example One of Two-stage Pressure Regulating Circuit   (b) Schematic Diagram of Electrical Control

Figure 4-2-2 Two-stage Pressure Regulating Circuit

1-Hydraulic Pump;2-Direct-acting overflow valve;3-Two-position three-way solenoid reversing valve;
4-Pilot type overflow valve;5-Shock-proof pressure gauge

Figure 4-2-3(a) is another example of two-stage pressure regulating circuit, where the piston drops to the working stroke and the high-pressure overflow valve 2 limits the maximum system pressure. The piston rises to non-working stroke and the low-pressure overflow valve 5 limits its maximum pressure. This circuit is commonly used in the hydraulic system of press machine.

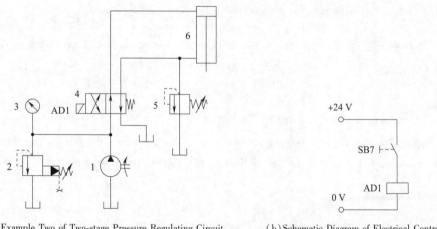

(a) Example Two of Two-stage Pressure Regulating Circuit   (b) Schematic Diagram of Electrical Control

Figure 4-2-3 Two-stage Pressure Regulating Circuit

1-Hydraulic pump;2-Pilot type overflow valve;3-Shock-proof pressure gauge;
4-Two-position four-way solenoid reversing valve;5-Direct-acting overflow valve;
6-Double-acting hydraulic cylinder with single piston rod

II Unloading Circuits

When hydraulic energy is not needed during work break of actuating component (or when work is stopped), the oil discharge from pump source shall be led to the oil tank directly in an automatic manner, in order to form an unloading circuit that enables the hydraulic pump to work under no-regulated and stabilized by overflow valve. In a variable pump system or a bypass throttle speed governing system, overflow valve (as safety valve) shall be used for limiting the maximum safety pressure of the system. When the system requires different working pressures in different working hours, two-stage or multi-stage pressure regulating circuit can be adopted.

1 Single-stage Pressure Regulating Circuit

As shown in Figure 4-2-1, the parallel connection of hydraulic pump 1 and overflow valve 2 forms a single-stage pressure regulating circuit. Regulation of the overflow valve pressure can change the output pressure of pump. After the regulation pressure of overflow valve is determined, the hydraulic pump works under the pressure regulated by the overflow valve. The pressure regulation and stabilization control of hydraulic system is thereby achieved.

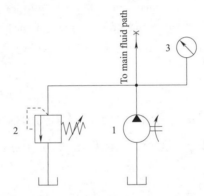

Figure 4-2-1 Single-stage Pressure Regulating Circuit
1-Hydraulic pump; 2-Direct-acting overflow valve; 3-Shock-proof pressure gauge

2 Two-stage Pressure Regulating Circuit

Figure 4-2-2(a) shows a typical two-stage pressure regulating circuit. This circuit is able to achieve two different system pressure controls, where the pilot type overflow valve 4 and the direct-acting overflow valve 2 regulate one stage, respectively. When the two-position three-way solenoid reversing valve 3 is in the position shown in the figure, the system pressure is regulated by valve 4. When valve 3 is energized and changes position, the system pressure is regulated by valve 2, where the overflow discharge of hydraulic pump returns to oil tank via the main valve 4, and valve 2 is also working and has oil flowing through.

Note: Regulation pressure of valve 2 must be less than the regulation pressure of valve 4.

The commutation circuits of solenoid reversing valves are most widely used, especially in hydraulic systems of combined machine tool with high requirements on automation degree. Such commutation circuits have been seen several times in many circuits above, so they are not described in more detail here. For scenarios having large flow and high requirement on commutation stability, commutation circuits of solenoid reversing valves cannot fulfill the requirements above anymore; and the commutation circuits using manually or mechanically operated reversing valve as pilot valve and using hydraulically operated reversing valve as main valve are usually adopted.

 Task implementation

1 Query through network to find out what the function of component 3 in Figure 4-1-2 is.
2 Draw the electrical control schematic diagram of Figure 4-1-2 through teamwork.
3 When the component 3 of the commutation circuit of the three-position four-way solenoid reversing valve in Figure 4-1-2 is de-energized, is the hydraulic cylinder in a locking or a floating state?

Task 2   Installation and Commissioning of Pressure Control Circuits

Task description

Analyze the patterns of commonly used pressure control circuits, and be able to tell the principles of commonly used pressure control circuits.

Relevant knowledge

I   Pressure Regulating and Limiting Circuits

Pressure control circuits use pressure valves to control and regulate the pressure of the main oil path or a certain branch of a hydraulic system, in order to fulfill the force or torque requirement for actuating component speed changeover circuit. Pressure control circuits can realize many controls of the system, such as pressure regulation (stabilization), depressurization, pressurization, unloading, pressure maintaining and balance. When the hydraulic system is working, the hydraulic pump shall provide the system with hydraulic fluid of required pressure, which saves energy, reduces the heat of fluid, and also improves the movement stability of actuating components. Therefore, pressure regulating or limiting circuits shall be provided. If the hydraulic pump works at the set pressure of system from time to time, the working pressure of hydraulic pump shall be

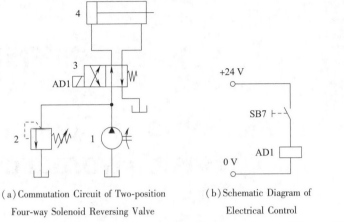

(a) Commutation Circuit of Two-position
Four-way Solenoid Reversing Valve

(b) Schematic Diagram of
Electrical Control

Figure 4-1-1　Commutation Circuit of Two-position Four-way Solenoid Reversing Valve

1-Hydraulic pump; 2-Direct overflow valve; 3-Two-position four-way solenoid reversing valve;
4-Double-acting hydraulic cylinder with single piston rod

II　Commutation Circuit of Three-position Four-way Solenoid Reversing Valve

As shown in Figure 4-1-2, when AD2 is energized, the pressure fluid of hydraulic pump 1 moves to the left chamber of hydraulic cylinder 4 via the right position of reversing valve 3, and the piston rod of hydraulic cylinder 4 extends. When AD2 is de-energized, AD1 is energized, the reversing valve 3 is in neutral position, and the hydraulic cylinder 4 stops moving. When AD1 is energized, the pressure fluid of hydraulic pump 1 moves to the right chamber of hydraulic cylinder 4 via the left position of reversing valve 3, and the piston rod of hydraulic cylinder 4 retracts.

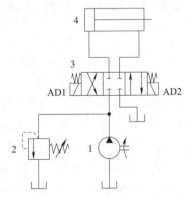

Figure 4-1-2　Commutation Circuit of Three-position Four-way Solenoid Reversing Valve

1-Hydraulic pump; 2-Direct overflow valve; 3-Three-position four-way solenoid reversing valve;
4-Double-acting hydraulic cylinder with single piston rod

Part 4

Installation and Commissioning of Basic Hydraulic Circuits

Task 1 Installation and Commissioning of Directional Control Circuits

 Task description

In a hydraulic system, there are many control means for commutation circuits; the objectives of task are to understand the core components and parts of commutation circuits, and to draw the electrical schematic diagrams of control circuits. The task further analyzes where the hydraulic cylinder is retracting or floating under different control conditions.

Relevant knowledge

I Commutation Circuit of Two-position Four-way Solenoid Reversing Valve

In general, commutation of moving parts can be achieved by various reversing valves. As shown in Figure 4-1-1, the left-right commutation of hydraulic cylinder is achieved by using a two-position four-way solenoid reversing valve.

When the reversing valve 3 is under the state as shown in Figure 4-1-1(a), the pressure fluid of hydraulic pump 1 moves to the left chamber of hydraulic cylinder 4 via the right position of reversing valve 3, and the piston rod of hydraulic cylinder 4 extends. After pressed the SB7 button, the electromagnet AD1 is energized, the pressure fluid of hydraulic pump 1 moves to the right chamber of hydraulic cylinder 4 via the left position of reversing valve 3, and the piston rod of hydraulic cylinder 4 retracts.

intended to reduce the fluid discharge resistance when the piston of the hydraulic cylinder returns to enable quick return and also extend the service life of the relief valve. In the clamping process, to prevent the fixture from damaging the workpiece, the clamping speed should be adjustable, so a one-way throttle valve is used in the circuit to adjust the extension speed of the hydraulic cylinder piston. The center-position unloading function of the reversing valve is to ensure that the drill bit will not act improperly due to obtaining sufficient pressure before the hydraulic cylinder clamps the workpiece. If the reversing valve in this circuit is changed to M-type center-position valve, when the clamping hydraulic cylinder is in the pressure-holding state, the pump is relieved of pressure, and the drilling cylinder cannot obtain sufficient pressure to work.

5 Discuss in groups (3-5 persons a group) and analyze the data and phenomena of the experiment. In the experiment, it can be observed that as fluid return throttle is used, the left and right chambers maintain high pressure when the hydraulic cylinder extends, and the pressure in the left chamber quickly rises to the set value when it extends into position, while the pressure in the right chamber slowly drops to zero. For the workpiece, the clamping force increases gradually, so that it is not easy to cause damage to the workpiece. However, pressure detection is not convenient for this pressure change process, which may even cause the pressure detector to act improperly.

To change the circuit to electrical control circuit and install a pressure relay in the left chamber of the hydraulic cylinder to automatically detect the clamping pressure, the throttling mode of the circuit must be changed to fluid-inlet throttling mode. As the pressure in the left chamber is very low when the fluid-inlet throttling hydraulic cylinder extends with no load, after the movable port contacts the workpiece, the pressure slowly rises to the maximum value. This ensures that the pressure relay acts reliably, without causing damage to the workpiece due to improper action.

Task implementation

1. Draw the graphic symbols of sandwich valves on equipment.
2. Establish a sandwich valve circuit.
3. Analyze the fault of a sandwich valve circuit and carry out simple troubleshooting.

Thinking and Exercise

1. What are the hydraulic control devices? What about their basic working principles?
2. Complete the circuit connection shown in the figure below on the equipment.
3. Discuss in groups and summarize the functions of different sandwich valve circuits in different application scenarios.

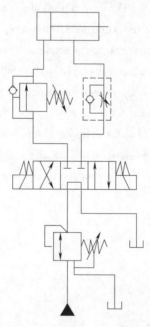

4. Analyze the functions of the circuit. The hydraulic circuit shall be designed by first considering that the clamping device can clamp workpieces made of different materials as the required clamping force differs with the workpiece material. In this circuit, one hydraulic pump is used to supply fluid for lifting a drill bit and clamping a workpiece. If an overflow valve is used to adjust the clamping pressure, the pressure will not be enough for the drill bit in drilling, so an overflow and relief valve is set in the circuit to adjust the clamping pressure.

This circuit used to reduce the working pressure of a branch or actuator in the circuit is called a pressure reducing circuit. A check valve connected in parallel to and beside the relief valve is

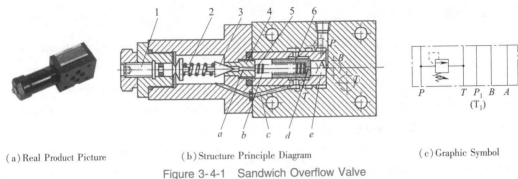

(a) Real Product Picture　　(b) Structure Principle Diagram　　(c) Graphic Symbol

Figure 3-4-1　Sandwich Overflow Valve

1-Push rod; 2-Spring; 3-Conical valve element; 4-Valve seat; 5-Spring; 6-Main valve element

II　Assembly of sandwich hydraulic system

Sandwich valve can be self-contained system. For sandwich valves of a diameter series, their main fluid path and the size, position and quantity of screw holes are the same as those of the plate type reversing valve of corresponding diameter. Therefore, an integrated hydraulic system can be assembled by stacking and directly connecting sandwich valves of the same diameter series.

Figure 3-4-2(a) shows the appearance of a sandwich hydraulic device. Its structure is shown in Figure 3-4-2(b). At its bottom is a base plate, which accommodates fluid inlet, fluid return port and fluid port leading to hydraulic actuator. The first component on the base plate is generally a pressure gauge switch, on which pressure control valves and flow control valves are stacked in order and a reversing valve is in the top layer. Bolts are used to fix them into a sandwich valve bank. Generally, one sandwich valve bank controls one actuator. If a hydraulic system has several hydraulic actuators that need to be controlled in a centralized manner, multiple base plates can be arranged side by side on it to form corresponding sandwich valve banks.

(a) Real Product Pictures of Two and Three Sandwich Valve Banks　　(b) Structure Diagram

Figure 3-4-2　Sandwich Hydraulic Device

Relevant knowledge

Sandwich valve is a hydraulic control valve that has developed rapidly in recent years. Compared with ordinary hydraulic valves, the sandwich valve has many significant advantages.

Sandwich valve is short for hydraulic sandwich valve. Its valve body is a component and also a connecting body with fluid path. The upper and lower sides of the valve body are made as connecting surfaces. By stacking sandwich valves of the same diameter series and fastening them with bolts, a desired hydraulic transmission system can be formed. Sandwich valves can be classified by the function into three types: pressure control valve, flow control valve, and directional control valve. Directional control valves of this type include check valves only, and main reversing valves are not sandwich valves.

I Structure and working principle of sandwich valve

The working principle of the sandwich valve is the same as that of general hydraulic valves, except for its specific structure. An overflow valve of this type is taken as an example here to describe its structure and working principle.

Figure 3-4-1(a) shows the appearance of a pilot-operated sandwich overflow valve. Its structure principle is shown in Figure 3-4-1(b). It is composed of a conical pilot valve and a main valve that is equivalent to a conical check valve. Its working principle is as follows: Pressure fluid flows from port P into the chamber e at the right end of the main valve element 6, and flows through the damping hole d on the valve element into the chamber b at the left end of the main valve element, and acts on the conical valve element 3 via the small hole a. When the system pressure is lower than the set pressure of the overflow valve, the conical valve is closed, the main valve is also closed, and the valve does not allow overflow. When the system pressure reaches the set pressure of the overflow valve, the conical valve element 3 opens, fluid in the chamber b flows back to the tank from the fluid port T via the conical valve port and the hole c, and fluid in the right chamber of the main valve element flows leftward via the damping hole d, creating pressure difference between both ends of the main valve element. This pressure difference allows the main valve element to overcome the pressure exerted by the spring 5 and move leftward, and the main valve port opens to allow overflow from the fluid port P to the fluid port T. By adjusting the pre-compression of the spring 2, the regulating pressure (i.e. overflow pressure) of the overflow valve and be adjusted. Figure 3-4-1(c) shows its graphic symbol.

Therefore:

$$p_2 - p_3 = \Delta p = F_s / A \qquad (3\text{-}2)$$

As the spring stiffness is low and the displacement of the relief valve element is small during operation, it can be considered that F_s basically remains unchanged. Therefore, the pressure difference $p_2 - p_3$ between both ends of the throttle valve basically remains unchanged; this ensures a stable flow of fluid through the throttle valve.

III Common Faults of Flow Control Valves and Troubleshooting Methods

Common faults of flow control valves and Troubleshooting methods see Table 3-3-1.

Table 3-3-1 Common Faults of Flow Control Valves and Troubleshooting Methods

Fault Symptom	Cause Analysis	Troubleshooting Method
No flow or very little flow	(1) The throttle port is blocked or the valve element is stuck (2) The fit clearance between the valve element and the valve hole is too large, with much leakage	(1) Check and clean it, replace fluid, improve fluid cleanliness, and repair the valve element (2) Check the wear and sealing conditions, and repair or replace the valve element
Unstable flow	(1) Foreign particles in the fluid adhere to the edge of the throttle port, resulting in decreased passage section and lower speed (2) The throttle valve has much leakage of fluid inside or outside it and the flow loss is high so that the flow required for the running speed cannot be guaranteed	(1) Disassemble and clean the throttle valve, remove dirt, and replace the fluid filter or fluid (2) Check the gap between the valve element and the valve body as well as their machining accuracy, repair or replace out-of-tolerance parts, and check the sealing condition of relevant connecting parts or replace sealing parts

Task implementation

1 Draw the graphic symbols of flow control valves on equipment.
2 Proficiently adjust the flow control valve to regulate the flow of fluid path.
3 Discuss in groups and summarize the application scenarios of flow control valves.

Task 4 Knowledge about Sandwich Valve

Task description

Understand the basic working principles and symbols of common sandwich valves, be acquainted with the types of sandwich valves, and be familiar with the applications of sandwich valves in different scenarios.

changes in load and temperature have a great impact on its flow control stability. Therefore, it is only suitable for hydraulic systems where the load is low, temperature does not change severely or speed stability requirement is not high.

II Speed Regulating Valve

Figure 3-3-2 shows a working principle diagram of a speed regulating valve. In terms of structure, a speed regulating valve is a combination of a throttle valve 2 and a fixed differential relief valve 1 connected in series before it. The pressure p_1 at the hydraulic pump outlet (i. e. inlet of the speed regulating valve) is basically unchanged under regulation by the overflow valve, while the pressure p_3 at the speed regulating valve outlet is determined by the hydraulic cylinder load F. When fluid flows through the relief valve, its pressure drops to p_2. The pressure p_2 acts on the chambers d and c of the relief valve via channels e and f. The pressure p_3 of fluid at the throttle valve outlet acts on the upper chamber b of the relief valve via the feedback channel a. When the relief valve element is in a certain equilibrium position under the action of spring force F_s and fluid pressures p_2 and p_3 (ignoring friction and hydraulic driving force, etc.), the following equation applies

$$p_2 A_1 + p_2 A_2 = p_3 A + F_s \tag{3-1}$$

Where: A, A_1 and A_2 are the effective areas of pressure fluid in chambers b, c and d acting on the valve element respectively, and $A = A_1 + A_2$.

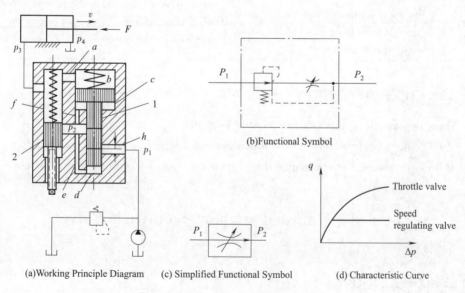

Figure 3-3-2 Speed regulating valve

1-Relief valve; 2-Throttle valve

Task 3　Knowledge about Flow Control Valve

Task description

Understand the basic working principles and symbols of common flow control valves, be acquainted with the types of flow control valves, and be familiar with the applications of flow control valves in different scenarios.

Relevant knowledge

A flow control valve regulates the output flow by changing the passage area of its port, thereby controlling the movement speed of an actuator. There are two types of commonly used flow control valves: throttle valve and speed regulating valve.

I　Throttle Valve

Figure 3-3-1 shows a structure principle diagram, graphic symbol and real product picture of an ordinary throttle valve. When the throttle valve is opened, pressure fluid flows in from the fluid inlet P_1 and flows out through hole a, axial triangular groove at the left end of the valve element 2, hole b and fluid outlet P_2. The valve element is always pressed against the end of the push rod 3 under the action of spring force. When the hand wheel 4 is rotated, the push rod moves in the axial direction to change the passage area of the throttle port, thereby regulating the flow through the valve.

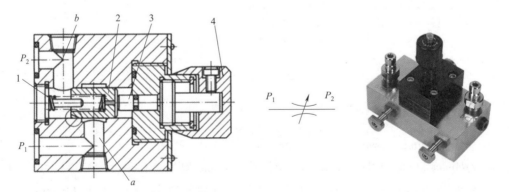

Figure 3-3-1　Ordinary Throttle Valve
1-Spring;2-Valve Element;3-Push Rod;4-Hand Wheel;a, b-Hole

A throttle valve is simple in structure, small in size, easy to use, and low in cost. However,

Continued

Fault Symptom	Cause Analysis	Troubleshooting Method
Large pressure fluctuation	(1) The main valve element does not act freely and sometimes gets stuck (2) The main valve element and the pilot valve seat damping holes are sometimes blocked and sometimes unblocked (3) The spring is bent or its stiffness is too small (4) The damping hole is too large and the vibration damping effect is poor (5) The pressure regulating nut is not locked	(1) Check, repair and reinstall it (apply tightening force of valve cover screw evenly), and filter or replace fluid (2) Clean the reduced damping hole and filter or replace fluid (3) Replace the spring (4) Reduce the damping hole appropriately (replace the valve element) (5) Tighten the pressure regulating nut after regulating pressure
High vibration amplitude and loud noise	(1) The radial force of the main valve element is unbalanced during operation, resulting in unstable performance of the overflow valve (2) Poor contact between the conical valve and the valve seat (excessive roundness error) results in unbalanced force exerted on and vibration of the conical valve (3) The pressure regulating spring is bent (or its axis is not perpendicular to the end face), resulting in unbalanced force exerted on and vibration of the conical valve (4) The flow passing through it exceeds the nominal flow, creating cavitation at the overflow valve port (5) The overflow flow passing through the overflow valve is too small, resulting in the overflow valve in the critical state of opening and closing and causing hydraulic shock	(1) Check the accuracy of the valve body hole and the main valve element, repair or replace parts, and filter or replace fluid (2) Control the fluid seal roundness error in the range of 0.005-0.01 mm (3) Replace the spring or grind the end face of the spring (4) Use it in the range of nominal flow only (5) Control the minimum overflow in normal operation

Task implementation

1 Draw the graphic symbols of pressure control valves on equipment.
2 Proficiently adjust the pressure control valve to regulate the pressure of fluid path.
3 Discuss in groups and summarize the application scenarios of pressure control valves.

Continued

Name	Overflow Valve	Relief valve	Sequence Valve
Fluid return characteristics	Fluid flows back to the tank from the valve, so the discharged fluid can be connected to the fluid return port in the valve body; this is internal discharging type	The fluid from the outlet of valve is secondary pressure fluid with pressure lower than the inlet fluid pressure and is supplied to the auxiliary fluid path, so a separate fluid discharge port is required; this is external discharge type	Like a relief valve, the fluid from the outlet is introduced to another cylinder, so a separate fluid discharge port is required; this is also external discharge type
Basic usage	It can be used as an overflow valve, safety valve or unloading valve and generally connected to the outlet of a pump and in parallel with the main fluid path. If it is used as a counterbalance valve, it needs to be connected in series to the fluid return path, with low set pressure	It is connected in series in the system and between a hydraulic pump and a branch fluid path	It is connected in series in the system and controls the sequential action of the actuator. It is mostly connected in parallel with check valve and used as one-way sequence valve
Example and description	For use as an overflow valve, the fluid path is normally open and the pump pressure depends on the regulated pressure of the overflow valve; this is mostly used in a fixed-displacement pump system for throttling and speed regulation. For use as a safety valve, the fluid path is normally closed and when the system pressure exceeds set value of the safety valve, the safety valve opens; this is mostly used in a variable displacement pump system	For pressure reduction, it allows the auxiliary fluid path to obtain pressure fluid with stable pressure lower than that of the main fluid path, and the valve port is normally open	It can be used as sequence valve and balance valve. The structure of sequence valve is similar to that of overflow valve; after proper modification, the two valves are interchangeable. However, the sequence valve requires high tightness; otherwise, it may cause improper action

Table 3-2-2   Common Faults of Pilot-operated Overflow Valves and Troubleshooting Methods

Fault Symptom	Cause Analysis	Troubleshooting Method
No pressure	(1) The main valve element damping hole is blocked (2) The main valve element is stuck in the opening position (3) The main valve balance spring is broken or bent, preventing the main valve element from returning to position (4) The pressure regulating spring is bent or omitted (5) The conical valve (or steel ball) is omitted or broken (6) The pilot valve seat is broken (7) The remote control port is connected to the tank	(1) Clean the damping hole and filter or replace fluid (2) Check, repair and reinstall it (apply tightening force of valve cover screw evenly), and filter or replace fluid (3) Replace the spring (4) Replace or install the spring (5) Install or replace it (6) Replace the valve seat (7) Check the working state of the solenoid reversing valve or the state (closed/open) of the remote control port

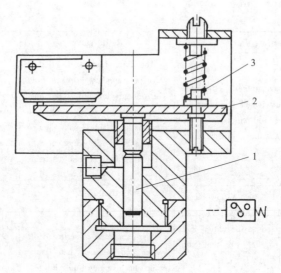

Figure 3-2-8 Plunger Type Pressure Relay
1-Plunger; 2-Lever; 3-Spring; 4-Switch

V Performance Comparison and Common Troubleshooting Methods for Pressure Control Valves

The structures and principles of various pressure control valves are very similar, with only partial differences in structure: some differ in the connection of fluid inlet and outlet, and some partially differ in the shape of valve element. Familiarity with the structure and performance characteristics of various pressure control valves is considerably conducive to analysis and troubleshooting of their faults. Table 3-2-1 lists the performance comparison of overflow valve, relief valve and sequence valve. Table 3-2-2 lists common faults of pilot-operated overflow valve and troubleshooting methods.

Table 3-2-1 Performance Comparison of Overflow Valve, Relief Valve and Sequence Valve

Name	Overflow Valve	Relief valve	Sequence Valve
Functional symbol			
Control fluid path Characteristics	Fluid is introduced from the fluid inlet to the bottom of the valve element to balance the spring force, so it controls the inlet fluid path pressure; the valve port is closed in normal conditions	Fluid is introduced from the fluid outlet to the bottom of the valve element to balance the spring force, so it controls the outlet fluid path pressure; the valve port is full open in normal conditions	Like the overflow valve, fluid is introduced from the fluid inlet to the bottom of the valve element, so it controls the inlet fluid path pressure; the valve port is closed in normal conditions

2 Application of sequence valve

Figure 3-2-7 shows a sequence valve used on a machine tool fixture to achieve a sequential action loop of positioning a workpiece before clamping. When the right position of the reversing valve U works, pressure fluid flows into the lower chamber of the positioning cylinder. After the positioning action is completed, the system pressure rises. When the pressure reaches the set pressure value of sequence valve (to ensure reliable work, the set pressure of sequence valve should be 0.5-0.8 MPa higher than the highest working pressure of the positioning cylinder), the sequence valve opens and pressure fluid flows into the lower chamber of the clamping cylinder through the sequence valve, achieving hydraulic clamping. When the left position of reversing valve works, pressure fluid flows into the upper chamber of the positioning cylinder and the clamping cylinder at the same time. When the positioning pin is pulled out and the workpiece is released, fluid returns through the check valve for the clamping cylinder. In addition, asequence valve can be also used as unloading valve, balance valve or counterbalance valve.

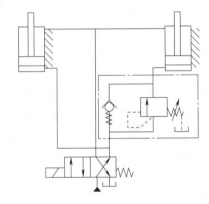

Figure 3-2-7 Application of sequence valve

IV Pressure Relay

A pressure relay is an electro-hydraulic control device that converts fluid pressure signal into electric signal. When fluid pressure reaches the set pressure of a pressure relay, it sends out electric signal to control the electromagnet, solenoid clutch, relay and other devices to act so as to depressurize or reverse fluid path, allow actuator to act sequentially, or turn off the motor to stop the system for safety protection. Figure 3-2-8 shows a structure diagram and functional symbol of a commonly used plunger type pressure relay. As shown in the figure, when the pressure of fluid introduced from the fluid inlet at the lower end of the pressure relay reaches the set pressure, the plunger 1 is pushed upward, and this displacement is amplified by the lever 2 to push the switch 4 to act. The acting pressure of the pressure relay can be adjusted by changing the compression of the spring 3.

an overflow valve leads to the tank. In addition, as there is pressure fluid at both fluid inlet and outlet of a sequence valve, its discharge port L must be connected to the tank separately. As the control fluid of a sequence valve is directly introduced from the fluid inlet, it is called an internal controlled and external discharging sequence valve.

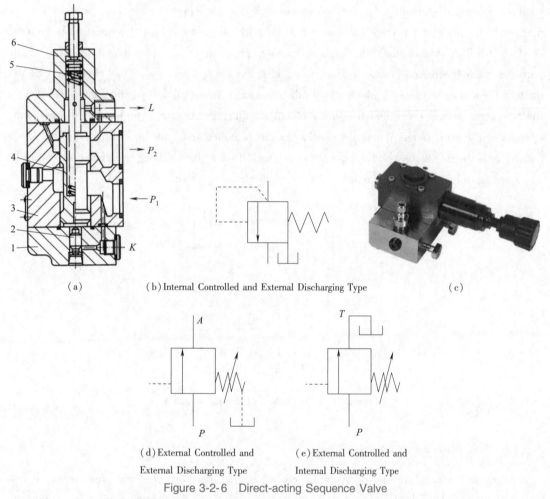

Figure 3-2-6 Direct-acting Sequence Valve
1-Lower cover; 2-Piston; 3-Valve body; 4-Valve element; 5-Spring; 6-Upper cover

By turning the lower cover 1 shown in Figure 3-2-6(a) by 90° or 180° in installation, cutting off the original fluid path, and removing the plug from the external control port K to connect the control fluid path, the valve would be opened by external pressure fluid and become an external controlled and external discharging sequence valve; its graphic symbol is shown in Figure 3-2-6(d). By further turning the upper cover 6 by 180° in installation and blocking the external discharge port L, the spring chamber would be connected to the fluid outlet and the valve becomes an external controlled and internal discharging sequence valve; its graphic symbol is shown in Figure 3-2-6(e).

path. Figure 3-2-5 shows a principle diagram of a relief valve used in clamping fluid path. In addition to supplying pressure fluid to the main fluid path, the hydraulic pump also supplies, through the relief valve on the branch fluid path, to the clamping cylinder stable pressure fluid with pressure lower than that supplied from a hydraulic pump. Its clamping force is adjusted and controlled by the relief valve.

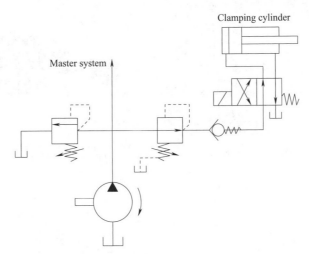

Figure 3-2-5 Application of relief valve

III Sequence Valve

A sequence valve controls the sequence of action of actuators in a hydraulic system. Sequence valves can be divided by the control pressure into two types: internal controlled and external controlled valves. The former controls the opening and closing of valve element by using inlet pressure of the valve, while the latter controls the opening and closing of valve element by using external control pressure fluid (i. e. hydraulic control sequence valve). Sequence valves are also available in direct-acting and pilot-operated valves. The former are generally used in low-pressure systems, while the latter are used in medium- and high-pressure systems.

1 Working principle of sequence valve

Figure 3-2-6 shows a working principle diagram and graphic symbol of a direct-acting sequence valve. When the fluid inlet pressure p_1 is low, the valve element is held at the lower end under the action of the spring; the fluid inlet and outlet are not connected. When the hydraulic pressure of fluid acting on the lower end of the valve element is greater than the preload of the spring, the valve element moves upward, the valve port opens, and fluid flows out from the fluid outlet through the valve port, thereby operating another actuator or other components to act. The figure indicates that the structures of a sequence valve and overflow valve are basically similar, except that the fluid outlet of a sequence valve leads to another pressure fluid path of the system, while the fluid outlet of

process can be deduced like the above-mentioned pilot-operated overflow valve, without repetition here.

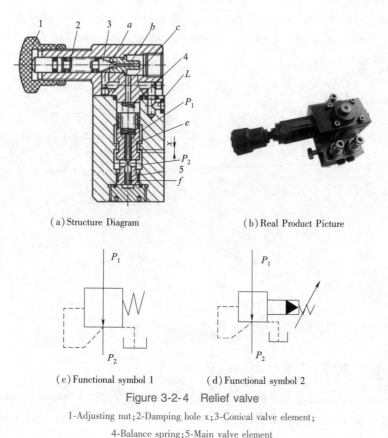

Figure 3-2-4　Relief valve
1-Adjusting nut;2-Damping hole x;3-Conical valve element;
4-Balance spring;5-Main valve element

Comparison of a pilot-operated relief valve and a pilot-operated overflow valve indicates the following differences between them:

①A relief valve keeps the outlet pressure basically unchanged, while an overflow valve keeps the inlet pressure basically unchanged.

②When not in operation, the fluid inlet and outlet of a relief valve are connected, but those of an overflow valve are not connected.

③To ensure that the set value of the outlet pressure of a relief valve is stable, its pilot valve spring chamber needs to be connected to the tank separately through the fluid discharge port. In comparison, the fluid outlet of an overflow valve is connected to the tank, so the spring chamber of its pilot valve and fluid to be discharged can be connected to the fluid outlet through the channel on the valve body, without need to be connected to the tank separately.

2　Application of relief valve

Relief valves are widely used in clamping fluid path, control fluid path and lubricating oil

II Relief Valve

A relief valve is a pressure control valve that enables the outlet pressure (secondary pressure) to be lower than the inlet pressure (primary pressure). Its function is to reduce the fluid pressure of a circuit in the hydraulic system, and provide two or several different pressure outputs with one fluid source at the same time. Relief valves are widely used in clamping systems, lubrication systems and control systems of various kinds of hydraulic equipment. In addition, when fluid pressure becomes unstable, a relief valve can be connected in series to the circuit to achieve stable low pressure. Relief valves are also available in direct-acting and pilot-operated valves. Pilot-operated relief valves are more widely used.

1 Working principle of relief valve

Figure 3-2-4(b) shows the appearance of a pilot-operated relief valve. Its structure principle is shown in Figure 3-2-4(a). It is similar to a pilot-operated overflow valve in structure, and also composed of two parts: pilot valve and main valve. Pressure fluid flows into the fluid inlet chamber P_1 from the fluid inlet of the valve, and after being depressurized by the relief valve with port x, flows out from the fluid outlet chamber P_2 and the fluid outlet. Pressure fluid from the fluid outlet chamber flows into the lower end of the main valve element 5 through the small hole f and meanwhile flows into the upper end of the main valve element through the small damping hole e, and acts on the conical valve element 3 through the holes c and b. When the outlet pressure is low, the pilot valve is closed, the pressures on both ends of the main valve element are equal; the main valve element is pressed to the lowest end (in the position shown in the figure) by the balance spring 4; the opening of the relief valve reaches the maximum extent and the pressure drop is the smallest; the relief valve does not have pressure reducing effect. When the fluid outlet pressure reaches the set pressure of the pilotvalve, the pilot valve opens. At this moment, part of pressure fluid in the chamber P_2 flows back to the tank through holes e, c and b, pilot valve hole a and discharge port L. Due to the effect of the small damping hole e, pressure difference is formed between both ends of the main valve element. Under the effect of this pressure difference, the main valve element overcomes the elastic force of the balance spring and rises; as a result, the relief valve port becomes smaller to reduce the fluid outlet pressure to the set value. If the fluid outlet pressure changes due to external interference (such as load change), the relief valve will automatically adjust its port opening to keep the fluid pressure stable. By adjusting the nut 1, the pre-compression of the pressure regulating spring 2 can be adjusted to regulate the fluid outlet pressure of the relief valve. The pressure regulating range of the medium-pressure pilot-operated relief valve is 2.5-8.0 MPa, which is also suitable for medium- and low-pressure systems. Figure 3-2-4(c) shows the graphic symbol of direct-acting relief valve; it is also the general symbol of relief valve. Figure 3-2-4(d) shows the graphic symbol of pilot-operated relief valve; its work

2 Functions of overflow valve

Based on function of overflow valve in a hydraulic system, the overflow valve can be also used as a safety valve, unloading valve or counterbalance valve in addition to realizing overflow.

(1) Acting as overflow valve.

In a hydraulic system where a fixed-displacement pump is used to supply fluid, the flow control valve regulates the flow of fluid into the actuator, and the excess fluid outputted by the fixed-displacement pump flows back to the tank from the overflow valve. In the working process, the overflow valve port is normally open, and the working pressure of the system is adjusted via the overflow valve and kept basically constant, as shown in Figure 3-2-3(a).

(2) Acting as safety valve.

Figure 3-2-3(b) shows a variable displacement pump based fluid supply system. The speed of the actuator is adjusted by the variable displacement pump, with no excess fluid in the system. The system working pressure changes with the load. During normal operation, the overflow valve port is closed. Once the system is overloaded, the overflow valve port opens immediately, allowing fluid to flow back to the tank so that the system pressure will no longer rise to ensure system safety.

(3) Acting as unloading valve.

As shown in Figure 3-2-3(c), the remote control port K of the pilot-operated overflow valve is connected to the tank via a two-position two-way solenoid valve. When the electromagnet is de-energized, the remote control port K is blocked, and the overflow valve functions for overflow and pressure stabilization. When the solenoid is energized, the remote control port K is connected to the tank, and the pressure at the upper end of the main valve element of overflow valve is close to zero. At this moment, the overflow valve port is fully opened, the fluid return resistance is small, and the fluid delivered from the pump flows back under low pressure to the tank through the overflow valve port to unload the hydraulic pump, thereby reducing the power loss of the system. Therefore, the overflow valve acts as unloading valve.

(4) Acting as counterbalance valve.

As shown in Figure 3-2-3(a), the overflow valve 2 is connected to the fluid return path; it can create resistance to fluid return and form back pressure to improve the motion smoothness of the actuator.

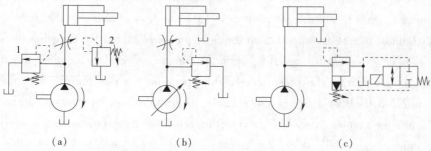

Figure 3-2-3 Functions of overflow valve

conical valve element 3 are equal. Under the action of the soft main valve spring 2, the valve element 3 is held at the lowest end, and the valve ports P and T of the overflow valve are separated, with no overflow. When the fluid inlet pressure rises to the extent that the hydraulic pressure acting on the pilot valve is greater than the acting force of the pilot valve spring, the pilot valve opens to allow pressure fluid to flow back to the tank through the damping hole and the pilot valve. Due to the damping hole, the hydraulic pressure p_2 at the upper end of the main valve element is less than the pressure p_1 at the lower end. When the force of this pressure difference acting on the main valve element becomes equal to or exceeds the preset spring force F_s of the main valve, the main valve element opens to allow fluid to flow in from port P and back to the tank via the main valve port T to achieve overflow.

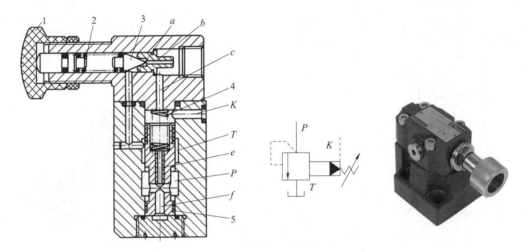

Figure 3-2-2　Pilot-operated overflow valve

1-Adjusting nut;2-Pressure regulating spring;3-Conical valve element;4-Spring;5-Main valve element

　　The pilot-operated overflow valve has a remote control port K. If port K is connected with a pipe to another remote pressure regulating valve (the remote pressure regulating valve has the same structure as the pilot control part of the overflow valve), the spring force of the remote pressure regulating valve can be adjusted to adjust the hydraulic pressure at the upper end of the main valve element of overflow valve so as to realize remote regulation of the overflow pressure of overflow valve. However, the maximum pressure that can be adjusted by the remote pressure regulating valve must not exceed the regulating pressure of the pilot valve of overflow valve. When the remote control port K is connected to the tank through a two-position two-way valve, the pressure at the upper end of the main valve element is close to zero. As the main valve spring is soft, the main valve element rises to the highest position, and the valve port opens widely. At this moment, the pressure at the overflow valve port P is very low; the system fluid flows back to the tank through the overflow valve under low pressure to realize unloading.

element through the damping hole a in the middle of the valve element 3. When the inlet fluid pressure is low, the valve element is held at the lower end under the action of the spring 2, separating port P from port T. When the fluid pressure increases, the force acting on the lower end of the valve element exceeds the compression force F of the spring. At this moment, the valve element rises, and the valve port is opened to discharge excess fluid back to the tank. The damping hole a in the valve element damps the action of the valve element to improve the balance of the valve. By adjusting the nut 1, compression force of the spring can be changed to adjust the fluid pressure p at the inlet of the overflow valve.

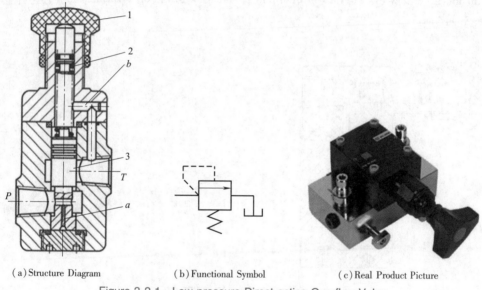

(a) Structure Diagram (b) Functional Symbol (c) Real Product Picture

Figure 3-2-1 Low-pressure Direct-acting Overflow Valve

1-Nut; 2-Pressure regulating spring; 3-Valve element

This low-pressure direct-acting overflow valve is generally used in small flow applications with pressure less than 2.5 MPa. Figure 3-2-1(b) shows the graphic symbol of a direct-acting overflow valve. Figure 3-2-1(a) reveals that in the normal position, the fluid inlet and outlet of the overflow valve are not connected, and the hydraulic pressure acting on the valve element is exerted by inlet fluid pressure. The fluid discharged via the overflow valve element flows into the fluid return port T through the internal discharge path.

(2) Pilot-operated overflow valve

Figure 3-2-2 shows a structure diagram of pilot-operated overflow valve. In the figure, pressure fluid flows in from port P and acts on the conical valve element 3 through the damping hole e. When the pressure at the fluid inlet is low, the hydraulic force acting on the pilot valve is not strong enough to overcome the acting force of the spring 2 on the right side of the pilot valve, the pilot valve is closed and no fluid flows through the damping hole, so the pressures on both ends of the

Task implementation

1. Draw the graphic symbols of directional control valves on equipment.
2. Proficiently adjust the directional control valve to change the direction of fluid path.
3. Discuss in groups and summarize the application scenarios of directional control valves.

Task 2　Knowledge about Pressure Control Valve

 Task description

Understand the basic working principles and symbols of common pressure control valves, be acquainted with the types of pressure control valves, and be familiar with the applications of pressure control valves in different scenarios.

 Relevant knowledge

In a hydraulic transmission system, the hydraulic valve controlling fluid pressure is called pressure control valve or pressure valve for short. The common point of these valves is that they operate under the principle of balancing hydraulic pressure and spring force acting on valve element.

Different hydraulic systems have different requirements for pressure control depending on the need of work: some are used to limit the maximum pressure of hydraulic system (e.g. safety valve); some are used to stabilize the pressure value (or pressure difference, pressure ratio, etc.) at a point in hydraulic system (e.g. overflow valve, relief valve and other constant pressure valves); others are controlled based on hydraulic pressure as controlling signal (e.g. sequence valve, pressure relay, etc.).

Ⅰ Overflow Valve

1 Structure and working principle of overflow valve

Commonly used overflow valves can be classified by the structure and basic action into two types: direct-acting and pilot-operated valves.

(1) Direct-acting overflow valve

A direct-acting overflow valve relies on pressure fluid in the system to directly act on the valve element and balance the spring force to control the valve element for opening and closing. Figure 3-2-1(a) shows a low-pressure direct-acting overflow valve, for which P is the fluid inlet, and T is the fluid return port. Pressure fluid at the inlet acts on the bottom end face of the valve

Table 3-1-4 Common Faults of Directional Valves and Troubleshooting Method

Fault Symptom	Cause Analysis	Troubleshooting Method
Valve element does not move or move into place	(1) The slide valve is stuck ①The clearance between the slide valve (valve element) and the valve body is too small so that the valve element is prone to be stuck in the hole and thus cannot act or act properly ②The valve element (or valve body) is damaged and the fluid is contaminated ③The valve element geometry is out of tolerance. The valve element and valve hole are not concentric, resulting in the element being stuck by axial hydraulic pressure (2) Control fluid path of hydraulic reversing valve is faulty ①The fluid control pressure is not enough, the slide valve does not act, allow reversing or reversing into place ②The throttle valve is closed or blocked ③The fluid discharge ports at both ends of the slide valve are not connected to the tank or the fluid discharge pipe is blocked (3) Electromagnet is faulty ①The AC electromagnet was burnt out due to the stuck slide valve or failure to attract the iron core to the bottom surface ②Magnetic flux leakage occurs or the attracting force is insufficient ③The electromagnet wiring is poorly welded and in poor contact (4) The spring is broken, omitted or too soft and cannot restore the slide valve to the center position, resulting in reversing failure The push rod of the solenoid reversing valve is worn so that its length is not enough or the stroke is incorrect so that the valve element moves insufficiently or excessively, resulting in reversing failure or not in place	(1) Check and repair the slide valve ①Check the clearance, and repair or replace the valve element ②Check, grind or re-fit the valve element, and replace fluid with new fluid if necessary ③Check and correct geometric deviation and concentricity, and check hydraulic sticking and repair it (2) Check the control fluid path ①Increase the control fluid pressure, and check whether the spring is too hard and replace it if it is too hard ②Check and clean throttle port ③Check and connect the fluid return tank, and clean the fluid return pipe to make it unblocked (3) Check and repair it ① Check whether the slide valve is stuck and replace the electromagnet ② Check the cause of magnetic flux leakage and replace the electromagnet ③Check and re-weld it (4) Check, replace or reinstall it Check and repair the push rod, and replace it if necessary
Reversing impact and noise	(1) The control flow is too high and the slide valve moves too fast, generating impact sound (2) The fit clearance between the one-way throttle valve element and the valve hole is too large, the spring of the check valve is omitted, and damping fails, generating impact sound (3) The contact surface of the electromagnet iron core is uneven or poorly connected (4) The slide valve jams and moves alternately or local friction is excessive (5) The bolt fixing the electromagnet is loose and causes vibration	(1) Reduce the throttle port of the one-way throttle valve and slow down the slide valve (2) Check and trim (repair) it to a reasonable clearance, and install the spring (3) Remove foreign objects and trim the electromagnet iron core (4) Grind, trim or replace the slide valve (5) Tighten the bolt and add a lock washer

position by positioning of the valve body. At this moment, the main valve ports P, A, B and T are not connected. In addition to the above-mentioned spring centering, hydraulic centering can also be used for electro-hydraulic reversing valves. For a hydraulic-centered electro-hydraulic reversing valve, when the pilot solenoid valve is in the center position, ports A' and B' are connected with port P, and port T'' is closed; the other aspects are similar to those of a spring-centered electro-hydraulic reversing valve.

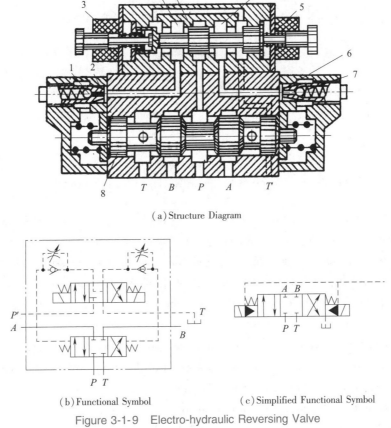

Figure 3-1-9　Electro-hydraulic Reversing Valve

1,6-Throttle valve; 2,7-Check valve; 3,5-Electromagnet; 4-Solenoid valve element; 8-Main valve element

4　Common faults of directional control valves and troubleshooting methods

Common faults of Directional Valves and Troubleshooting method see Table 3-1-4.

to allow fluid to pass at a high flow. In this way, a small electromagnet can be used to control high fluid flow.

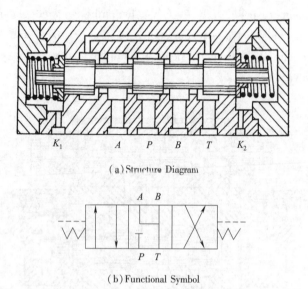

(a) Structure Diagram

(b) Functional Symbol

Figure 3-1-8 Three-position Four-way Hydraulic Reversing Valve

Figure 3-1-9 shows the structure and functional symbol of a spring-centered three-position four-way electro-hydraulic reversing valve. When the electromagnet on the left side of the pilot solenoid valve is energized, its valve element moves to rightward, and control pressure fluid from the main valve port P or external port flows via the pilot solenoid valve port A' and the left check valve into the chamber at the left end of the main valve, and pushes the main valve element to move rightward. At this moment, control fluid in the chamber at the right end of the main valve element flows via the right throttle valve, pilot solenoid valve ports B' and T', and main valve port T or external port and returns to the tank (the moving speed of the main valve element can be adjusted with the right throttle valve) to connect the fluid paths between the main valve ports P and A, and between ports B and T. When the electromagnet on the right side of the pilot solenoid valve is energized, the fluid paths between the ports P and B andbetween ports A and T can be connected. When the two electromagnets of the pilot solenoid valve are not energized, the pilot solenoid valve element returns to the center position under the action of its centering spring. At this moment, control pressure fluid from the main valve port P or the external port no longer flows into the left and right chambers of the main valve element. The left and right chambers of the main valve element is connected with the pilot solenoid valve port T' via the pilot solenoid valve ports A' and B' in the center (as shown in Figure 5-1-9(b)) and the fluid in the two chambers flows back to the tank through them and via the main valve port T of or the external port. The main valve element is driven by pre-pressure applied by the centering springs at both ends and accurately returns to the center

means of a spring. A three-position solenoid valve has two electromagnets. Figure 3-1-7 shows the structure and functional symbol of a three-position five-way solenoid reversing valve.

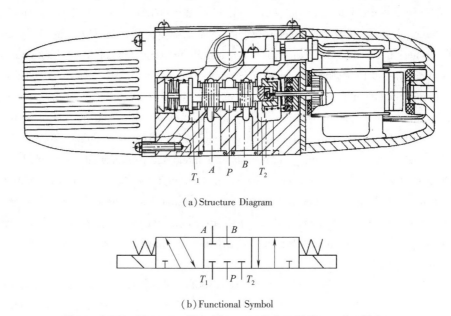

(a) Structure Diagram

(b) Functional Symbol

Figure 3-1-7　Three-position Five-way Solenoid Reversing Valve

(4) Hydraulic reversing valve: A hydraulic reversing valve changes the position of its valve element by using pressure fluid in the control fluid paths. Figure 3-1-8 shows the structure and functional symbol of a three-position four-way hydraulic reversing valve. The valve element is moved by the pressure difference of fluid in the sealed chambers at both ends. When pressure fluid of control fluid paths flows into the right chamber of the slide valve from the control port K_2 on the right of the valve, K_1 becomes connected to return fluid and the valve element moves leftward to connect pressure fluid ports P and B, and connect ports A and T. When K_1 is connected to pressure fluid and K_2 is connected to return fluid, the valve element moves rightward to connect ports P and A, and connect ports B and T. When both K_1 and K_2 are connected to return fluid, the valve element returns to the center position under the action of the springs at both ends and the locating sleeve.

(5) Electro-hydraulic reversing valve: In large- and medium-sized hydraulic equipment, when high flow passes through a valve, the friction and hydraulic force acting on the slide valve are great. In this case, the electromagnet thrust of a solenoid reversing valve is relatively too small, and the solenoid reversing valve should be replaced with an electro-hydraulic reversing valve. An electro-hydraulic reversing valve is a combination of solenoid slide valve and hydraulic slide valve. The solenoid slide valve functions as a pilot; it can change the direction of control fluid flow, thereby changing the position of the hydraulic slide valve element. As the hydraulic thrust for operating the hydraulic slide valve can be very strong, the size of the main valve element can be made very large

fast engagement and disengagement with action time of about 0.01-0.03 s. Its disadvantage is that if the supply voltage drops by more than 15%, the electromagnet attracting force is significantly reduced; if the armature does not act, the coil of a dry electromagnet will be burned out 10-15 minute later (1-1.5 h for a wet electromagnet); it also has considerable impact, loud noise, and short service life. Therefore, the allowable switching frequency of an AC electromagnet in practice is generally 10 times/min and not more than 30 times/min. DC electromagnet works more reliably, its engagement and disengagement action time is in the range of about 0.05-0.08 s, and its allowable switching frequency is higher, generally 120 times/min, and 300 times/min at most. It has small impact, small size, and long service life. However, it needs a special DC power supply, with high cost. In addition, there is also an integral electromagnet requiring DC current. The electromagnet is provided with a rectifier that converts inputted AC current to DC current before supplying it to the DC electromagnet. At present, an oil-immersed electromagnet has been newly developed abroad. Its armature and excitation coil are immersed in oil during operation. It features longer service life and operation with higher stability and reliability. However, due to the high cost, it is not widely applied.

Figure 3-1-6(a) shows the structure of a two-position three-way AC solenoid reversing valve. In the position shown in the figure, ports P and A are connected and port B is disconnected. When the electromagnet is engaged upon energization, the push rod 1 pushes the valve element 2 to the right and meanwhile, the fluid port P becomes disconnected from port A and connected with port B. When the electromagnet is de-energized and disengaged, the spring 3 pushes the valve element to reset. Figure 3-1-6(b) shows its functional symbol.

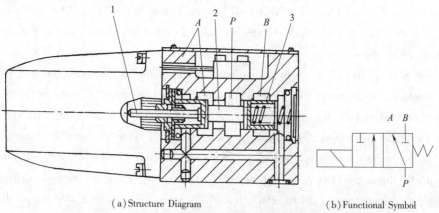

(a) Structure Diagram   (b) Functional Symbol

Figure 3-1-6   Two-position Three-way Solenoid Reversing Valve

1-Push rod; 2-Valve element; 3-Spring

As mentioned above, solenoid reversing valves are available with two positions and three positions in terms of working position. A two-position solenoid valve has one electromagnet reset by

(2) Mechanical reversing valve: A mechanical reversing valve, also known as a stroke valve, is mainly used to control the stroke of mechanical moving parts. Its valve element is forced to move by means of the stop iron 5 installed on the table to control the fluid flow direction. Mechanical reversing valves are generally available as two-position two-way, three-way, four-way and five-way valves. Two-position two-way mechanical valves are available in two types: normally closed and normally open valves. Figure 3-1-5(a) shows a roller type two-position three-way normally closed mechanical reversing valve. In the position shown in the figure, the valve element 2 is pressed by the spring 1 to the upper end, the fluid ports P and A are connected, and port B is closed. When the stop iron 5 presses the roller 4 to move the valve element 2 to the lower end, the fluid ports P and A are disconnected, ports P and B are connected and port A is closed. Figure 3-1-5(b) shows its functional symbol.

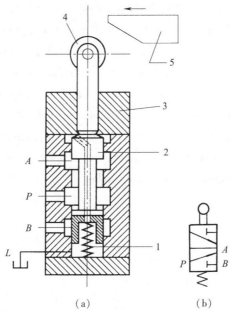

Figure 3-1-5 Mechanical Reversing Valve
1-Spring; 2-Valve element; 3-Valve body; 4-Roller; 5-Stop iron

(3) Solenoid reversing valve: A solenoid reversing valve controls fluid flow direction by using the electromagnet action of engagement upon energization and disengagement upon de-energization to directly push the valve element. It is a signal conversion element functioning between electrical system and hydraulic system.

Electromagnets can be divided into AC and DC electromagnets by the power supply, and into "wet type" and "dry type" electromagnets by presence of oil in the armature working chamber or not. An AC electromagnet features a large starting force without need of special power supply, and

(3) Starting smoothness: When the valve element is in the center position, if a chamber of the hydraulic cylinder is connected to the tank, starting will not be smooth enough as the chamber has no fluid for buffering in starting.

(4) Hydraulic cylinder "floating" and stopping at any position: When the valve element is in the center position and ports A and B are connected, the horizontal hydraulic cylinder "floats", and other mechanisms can be used to move the table and adjust its position. When ports A and B are blocked or connected with port P (in a non-differential condition), the hydraulic cylinder can be stopped in any position. The functions of a three-position five-way reversing valve are similar to those mentioned above.

3 Structure of reversing valve

Slide valve type reversing valves are widely used in hydraulic transmission systems. Some typical structures of these reversing valves are mainly introduced below.

(1) Manual reversing valve: Figure 3-1-4(a) shows an automatic resetting manual reversing valve. When the handle 1 is released, the valve element 2 automatically returns to the center position under the action of the spring 3. This valve is suitable for scenarios where the valve would act frequently for short duration each time, with relatively complete operation. The valve is commonly used in the hydraulic transmission system of construction machinery.

If the spring 3 at the right end of the valve element is changed to a structure that allows automatic positioning, the valve becomes a manual reversing valve that can be positioned in three positions. Figure 3-1-4(b) shows its functional symbol.

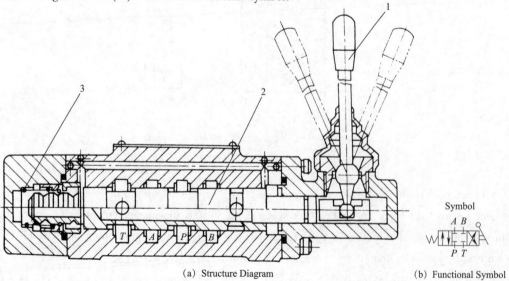

(a) Structure Diagram   (b) Functional Symbol

Figure 3-1-4 Manual Reversing Valve

1-Handle; 2-Valve element; 3-Spring

Table 3-1-3 List of Five Common Center-position Functions of
Three-position Four-way Reversing Valves

Code	Structural Diagram	Symbol of Center Position	Status and Characteristics of Ports in Center-position Condition
O			All fluid ports are fully enclosed. The reversing accuracy is high but with impact, the cylinder is locked, the pump is not unloaded, and the parallel cylinders can move
H			All fluid ports are fully open. Reversing is smooth, the cylinder floats, the pump is unloaded, and other cylinders cannot be used in parallel
Y			Port P is closed, and ports A, B and T are connected. Reversing is relatively stable, the cylinder floats, the pump is not unloaded, and the parallel cylinders can move
P			Port T is closed, and ports P, A and B are connected. Reversing is the smoothest, the double-rod cylinder floats, the single-rod cylinder is differential, the pump is not unloaded, the parallel cylinders can move
M			Ports P and T ports are connected, and Ports A and B are closed. The reversing accuracy is high but with impact, the cylinder is locked, the pump is unloaded and other cylinders cannot be used in parallel.

In analysis and selection of the center-position function of a valve, the following points should be considered in general:

(1) System pressure maintaining: When port P is blocked, the system maintains pressure and the hydraulic pump can be used in a multi-cylinder system. When port P is connected to port T in non-smooth way, the system can maintain a certain pressure for the control fluid path.

(2) System unloading: When port P is connected to port T, the system is unloaded.

Table 3-1-2 Brief description of the structure principle, graphic symbol and application scenarios of reversing valve

Name	Structure Principle Diagram	Graphic Symbol	Application Scenario
Two-position two-way reversing valve			It connects fluid and allows/stops fluid flow (equivalent to a switch)
Two-position three-way reversing valve			It controls fluid flow direction (from one direction to another)
Two-position four-way reversing valve			It does not allow an actuator to stop in any position
Two-position five-way reversing valve			It does not allow an actuator to stop in any position
Three-position four-way reversing valve			It allows an actuator to stop in any position
Three-position five-way reversing valve			It allows an actuator to stop in any position

2 Center-position function of reversing valve

When a reversing valve is in the normal position, the connection mode of its fluid ports enables the slide valve function. A three-position reversing valve is normally in center position, so the slide valve function of a three-position reversing valve is also known as center-position function. Three-position reversing valves with different center-position functions have common valve bodies and differ in the structure and size of spool land and internal through hole, with significant difference in functions. Table 3-1-3 shows list of five common center-position functions of three-position four-way reversing valves.

II Reversing Valve

A reversing valve allows or stops fluid flow or changes fluid flow direction in fluid paths based on its valve element motion relative to its valve body to start, stop or change movement direction of a hydraulic actuator. Table 3-1-1 shows the classification of reversing valves.

Table 3-1-1 Classification of Reversing Valves

Classification	Type
By form of motion of valve element	Slide valve, rotary valve
By working positions and ways of valve	Two-position two-way, two-position three-way, two-position four-way, three-position four-way, three-position five-way valves, etc.
By control mode of valve	Manual, mechanical, electrical, hydraulic, electro-hydraulic

1 Working principle of reversing valve

Figure 3-1-3 shows the working principle of reversing valve. In the state shown in the figure, the two chambers of the hydraulic cylinder do not allow pressure fluid flow, and the piston is stationary. If the valve element 1 is moved to the left, fluid ports P and A of the valve body 2 are connected, ports B and T are connected, pressure fluid flows into the left chamber of the hydraulic cylinder through ports P and A, and fluid in the right chamber flows back to the fluid tank through ports B and T to move the piston rightward. If the valve element is moved to the right, the ports P and B are connected, and ports A and T are connected to allow the piston to move leftward.

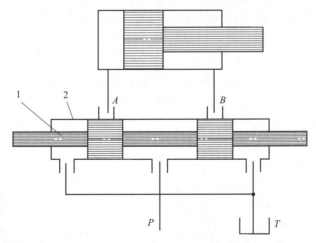

Figure 3-1-3 Working principle of reversing valve
1-Valve element; 2-Valve body

Brief description of the structure principle, graphic symbol and application scenarios of reversing valve see Table 3-1-2.

1 Ordinary check valve

An ordinary check valve functions to enable fluid to flow in only one direction and prevent it from flowing backward. Figure 3-1-1(a) shows the structure of a tubular ordinary check valve. When pressure fluid flows in from port P_1 at the left end of the valve body, it overcomes the force exerted by the spring 3 on the valve element 2 and moves the valve element rightward to open the valve port, and flows out from the port at the right end of the valve body through the radial hole a and the axial hole b on the valve element 2. But when pressure fluid flows in from port P_2 at the right end of the valve body, it and the spring press the conical surface of the valve element firmly against the valve seat to keep the valve port closed, preventing fluid from flowing through. Figure 3-1-1(b) shows the functional symbol of a check valve, and Figure 3-1-1(c) shows a real product picture.

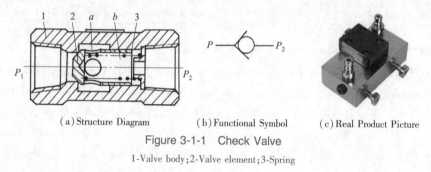

(a) Structure Diagram   (b) Functional Symbol   (c) Real Product Picture

Figure 3-1-1 Check Valve

1-Valve body; 2-Valve element; 3-Spring

2 Hydraulic control check valve

Figure 3-1-2(a) shows the structure of a hydraulic control check valve. When control port K has no pressure fluid flowing in, its working mechanism is the same as an ordinary check valve: pressure fluid can only flow from port P_1 to port P_2 and cannot flow reversely. When control port K has control pressure fluid, as the chamber a on the right side of the control piston 1 has a port that allows fluid to pass, the piston 1 moves rightward and pushes the ejector rod 2 to open the valve element 3; as a result, the ports P_1 and P_2 are connected, allowing fluid to flow freely in the two directions. Figure 3-1-2(b) shows the functional symbol of a hydraulic control check valve, and Figure 3-1-2(c) shows a real product picture.

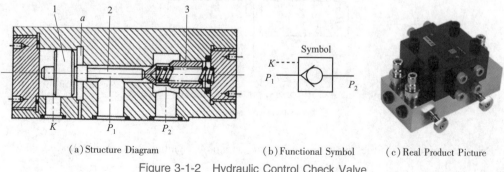

(a) Structure Diagram   (b) Functional Symbol   (c) Real Product Picture

Figure 3-1-2 Hydraulic Control Check Valve

1-Piston; 2-Ejector rod; 3-Valve element

Part 3
Hydraulic Control Valve

Task 1 Knowledge about Directional Control Valve

Task description

Understand the basic working principles and symbols of common directional control valves, be acquainted with the types of directional control valves, and be familiar with the applications of directional control valves in different scenarios.

Relevant knowledge

A hydraulic valve is used to control the flow direction of or adjust the pressure and flow rate of fluid in a hydraulic system. Therefore, it is available in three categories: directional valve, pressure valve and flow valve. Valves in the same shape may have different functions due to different mechanisms of action. Pressure valves and flow valves control the pressure and flow of a system based on the throttling effect of passage cross section, while directional valves control the flow direction of fluid by changing the fluid passage channel. In other words, despite various types of hydraulic valves, they still maintain some basic common points. For example:

In terms of structure, all these valves are composed of valve body, valve element (rotary valve or slide valve), and component actuating the valve element (e. g. spring, electromagnet).

In terms of working principle, the relationship between the opening size, valve inlet/outlet pressure difference, and through-valve flow of all these valves conforms to the orifice flow formula, although different types of valves differ in control parameters.

Directional control valves are used to allow or stop fluid flow in fluid paths in a hydraulic system, including check valve, reversing valve, pressure gauge switch, etc.

I Check Valve

Common check valves in hydraulic system include ordinary check valve and hydraulic control check valve.

the structural characteristics of the main machine.

For example, it is the simplest and most convenient to directly use a hydraulic cylinder to realize the reciprocating linear motion of the machine. For a scenario requiring consistent reciprocating motions, a hydraulic cylinder with double piston rods may be used. If quick return is required, a hydraulic cylinder with single piston rod should be used and differential connection considered. In the case of long stroke, a plunger cylinder may be used to reduce the difficulty of machining. In case of long stroke with a light load, some transmission devices may also be considered to expand the stroke. Reciprocating swinging motion may be realized either by a rotary cylinder, or by a linear hydraulic cylinder plus a connecting rod mechanism or gear-rack mechanism.

 Task implementation

1　Draw the graphic symbols of hydraulic cylinders on the equipment.

2　Proficiently design the bore diameter of hydraulic cylinder.

3　Discuss in groups and summarize the application of different hydraulic cylinders in different scenarios.

Thinking and Exercise

1　What are the hydraulic actuators? What about their basic working principles?

2　What about the relation of and difference between a hydraulic motor and a hydraulic pump?

3　Briefly describe the characteristics and applications of hydraulic cylinders.

3 Precautions during the design of hydraulic cylinder

Whether the hydraulic cylinder is designed and used correctly or not directly affects its performance and the probability of its failure. In this respect, the common problems are improper installation of hydraulic cylinder, sagging of hydraulic cylinder or piston, eccentric load of piston rod, and instability of compression bar of piston rod. Therefore, during the design of hydraulic cylinder, attention should be paid to the following:

(1) The piston rod of the hydraulic cylinder should bear the maximum load in tension state or have good stability in compression state as far as possible.

(2) Both braking at the stroke end of the hydraulic cylinder and venting of the hydraulic cylinder should be considered. In case of no cushion device and venting device in the cylinder, corresponding measures should be taken for the system. However, not all hydraulic cylinders involve this issue.

(3) The hydraulic cylinder should be installed and fixed correctly. If the piston rod subjected to bending needs rabbet connection rather than threaded connection, the hydraulic cylinder should only be positioned at one end, not positioned at both ends with keys or pins, so as not to hinder its expansion when heated. If the piston rod is compressed by the impact load, the positioning piece should be arranged at the piston rod end; otherwise, it should be arranged at the cylinder head end.

(4) The structure of each part of the hydraulic cylinder should be designed according to the recommended structural type and design standard, and its structure should be as simple and compact as possible for easy machining, assembly and maintenance.

(5) The outline dimensions of the hydraulic cylinder should be reduced as much as possible under the precondition that the motion stroke and load force are met.

(6) Reliable sealing performance and good dustproof performance of the hydraulic cylinder should be ensured. Reliable sealing performance is an important indicator of the normal work of the hydraulic cylinder. In case of any serious leakage, the working efficiency of the hydraulic cylinder may be reduced, or even its abnormal operation (such as the failing in meeting the requirements for load force and movement speed, etc.) may be caused. Good dustproof measures are helpful to prolong the service life of the hydraulic cylinder.

To sum up, the design contents of hydraulic cylinder are variable and they can be modified slightly based on the actual situation or some new contents may be added. The design steps should be modified repeatedly before getting correct and reasonable design results. For the design of hydraulic cylinder, selecting the correct type of hydraulic cylinder is the premise of all design calculations. During type selection of hydraulic cylinder, it is necessary to start from the requirements for stroke length, motion characteristics and motion performance of the machinery and equipment, as well as consider the installation space and specific installation position provided by

outer diameter of piston rod d and the length of barrel L.

(1) Inner diameter of barrel D. Determining the inner diameter of barrel of the hydraulic cylinder is to select the working pressure and the speed ratio of reciprocating motion based on the load, thus calculating the effective working area of the hydraulic cylinder and obtaining the inner diameter of barrel D. Next, the closest standard value given in GB/T 2348—2018 should be selected as the designed inner diameter of barrel.

D is determined according to the load and working pressure.

①If the chamber without rod serves as the working chamber

$$D=\sqrt{\frac{4F_{max}}{\pi p_i}} \qquad (2\text{-}23)$$

②If the chamber with rod serves as the working chamber

$$D=\sqrt{\frac{4F_{max}}{\pi p_i}+d^2} \qquad (2\text{-}24)$$

Where: p_i is the working pressure of the cylinder's working chamber, which can be determined based on the type or load of the machine tool; F_{max} is the maximum acting load.

(2) Outer diameter of piston rod d. Generally, the outer diameter of piston rod d should meet the speed or speed ratio requirement before the structural strength and stability are checked. If the speed ratio is recorded as λ_v, the following formula with a radical sign can be obtained:

$$D=\sqrt{\frac{\lambda_v-1}{\lambda_v}} \qquad (2\text{-}25)$$

The outer diameter of piston rod can also be determined according to the stress on it. Generally, when the piston rod bears a tensile stress, $d=(0.3\text{-}0.5)D$.

When the piston rod bears a compressive stress:

$d=(0.5\text{-}0.55)D$ if $p_i<5$ MPa;

$d=(0.6\text{-}0.7)D$ if 5 MPa$<p_i<7$ MPa;

$d=0.7D$ if $p_i>7$ MPa.

(3) Length of barrel L. The length of barrel L is determined based on the maximum working stroke plus various structural demands, namely

$$L=l+B+A+M+C$$

Where: l is the maximum working stroke of piston; B is the width of piston, generally taken as $(0.6\text{-}1)D$; A is the guide length of piston rod, take as $(0.6\text{-}1.5)D$; M is the sealing length of piston rod, which is determined by the sealing method; C is the length of other part.

Generally, the length of barrel should not exceed 20 times of its inner diameter.

In addition, structural dimensions of the hydraulic cylinder also include the minimum guide length H.

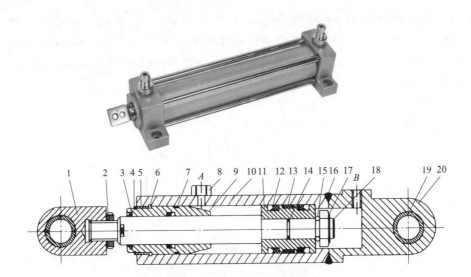

Figure 2-2-8 Double-Acting Hydraulic Cylinder with Single Piston Rod

1-Trunnion; 2-Nut; 3-Dust ring; 4&17-Spring collar; 5-Sleeve; 6&15-Stirrup;
7&14-O-ring; 8&12-Y-shaped seal ring; 9-Cylinder head and guide sleeve; 10-Barrel;
11-Piston; 13-Wear ring; 16-Stirrup cap; 18-Piston rod; 19-Bushing; 20-Cylinder bottom

IV Design of Hydraulic Cylinder

As an actuator of hydraulic transmission, hydraulic cylinder is directly related to the working mechanism of the main machine. For different machine types and mechanisms, hydraulic cylinder has corresponding applications and working requirements. Therefore, before designing the hydraulic cylinder, it is necessary to first analyze the working conditions of the entire hydraulic system, prepare the load diagram, select the working pressure for the system, then select the structural type as per the actual application requirements and determine the main working dimensions of the hydraulic cylinder according to the movement requirements, load conditions, maximum stroke, etc., and finally carry out the structural design after performing checking calculation for strength, stability and cushion.

1 Contents of and steps for design of hydraulic cylinder

(1) Select the type of hydraulic cylinder and the structural types of other parts.

(2) Determine working parameters and structural dimensions of the hydraulic cylinder.

(3) Calculate and check the structural strength and rigidity.

(4) Design the guiding, sealing, dust-proofing, venting and cushion devices.

(5) Plot the assembly drawing and the part drawing and compile the design specification.

2 Calculation of structural dimensions of hydraulic cylinder

Structural dimensions of a hydraulic cylinder mainly include the inner diameter of barrel D, the

(3) Gear cylinder is shown in Figure 2-2-7. It consists of two plunger cylinders and a set of rack and pinion transmission device. The movement of plunger is converted into the drive of gear via the rack and pinion transmission device to realize the reciprocating swing or intermittent feed motion of the workpiece.

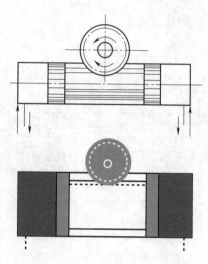

Figure 2-2-7   Gear Cylinder

III Typical Structures of Hydraulic Cylinders

Figure 2-2-8 shows a common double-acting hydraulic cylinder with single piston rod. It consists of cylinder bottom 20, cylinder head and guide sleeve 9, barrel 10, piston 11 and piston rod 18. One end of the barrel is welded to the cylinder bottom, and the other end of the barrel is fixed to the cylinder head (guide sleeve) with stirrup 6, sleeve 5 and spring collar 4 for easy removal, installation and maintenance. Fluid ports A and B are provided at both ends. Piston 11 is connected to piston rod 18 with stirrup 15, stirrup cap 16 and spring collar 17. The piston and the cylinder bore are sealed with a pair of Y-shaped polyurethane seal rings 12. Due to the certain gap between the piston and the cylinder bore, a wear ring (13, also called support ring) made of nylon 1010 is used for centering and guiding. The inner bore of piston rod 18 and piston 11 is sealed with a seal O-ring 14. A long guide sleeve 9 can ensure no decentralization of the piston rod. As for the guide sleeve, its outer bore is sealed with an O-ring 7 and its inner bore is sealed with a Y-shaped seal ring 8 to prevent fluid leakage. The use of dust ring 3 can effectively prevent dust from entering the cylinder. The cylinder and the rod end are connected to external parts through pin holes, in which nylon bushings are provided for wear resistance.

Obviously, the supercharging capability is obtained on the basis of reducing the effective energy. Namely, the supercharging cylinder only increases the output pressure and does not increase the output energy.

When the plunger moves to the end point, the single-acting supercharging cylinder does not output high-pressure fluid any more until the piston retracts to the left end and moves rightward. To make up for this shortcoming, a double-acting supercharging cylinder shown in Figure 2-2-5(b) may be used and fluid can be continuously supplied to the system via both high-pressure ends.

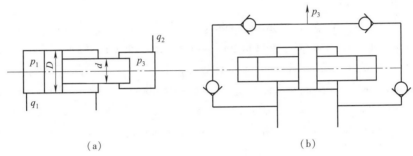

Figure 2-2-5 Supercharging Cylinder

(2) Telescopic cylinder. A telescopic cylinder consists of two or more piston cylinders, with the inner bore of piston rod of the previous-stage piston cylinder being the barrel of the next-state piston cylinder. A long working stroke can be obtained with the barrel extended and a small structural size kept with the barrel retracted. Telescopic cylinder is widely applied to lifting and transport vehicles.

The telescopic cylinder may be of single-acting type shown in Figure 2-2-6(a) or double-acting type shown in Figure 2-2-6(b). The former retracts depending on external force, while the latter retracts depending on hydraulic pressure.

Figure 2-2-6 Telescopic Cylinder

The extending movement of a telescopic cylinder is carried out stage by stage. The barrel with the maximum diameter starts to extend at the minimum fluid pressure first; when this barrel moves to the end point of its stroke, the barrel with a smaller diameter starts to extend; the last-stage barrel with the minimum diameter extends at last. With the increase in the serial number of working stage, the diameter of extending barrel becomes smaller, resulting in pressure rise of the working fluid and increase of the operating speed.

$$F_3 = p_1(A_1 - A_2) = p_1 \pi d^2 / 4 \quad (2\text{-}18)$$
$$v_3 = 4q / \pi d^2 \quad (2\text{-}19)$$

Formulae (2-18) and (2-19) show that the hydraulic cylinder with differential connection has a lower thrust and higher speed than that without differential connection. By taking advantage of this, a faster movement speed can be obtained without increasing the flow rate of fluid source. This connection mode is widely applied to the rapid movement of the hydraulic power system of combined machine tool and other mechanical equipment.

2 Plunger cylinder

Figure 2-2-4(a) shows a plunger cylinder, which can realize unidirectional hydraulic transmission only. The reverse movement of this cylinder depends on external force. If bidirectional movement is required, a pair of plunger cylinders must be used. As shown in Figure 2-2-4(b), without the contact between plunger and barrel of the hydraulic cylinder, the movement is guided by the guide sleeve on the cylinder head. Therefore, this type of cylinder is especially applicable to the long stroke scenario without finish machining of the inner wall of the barrel.

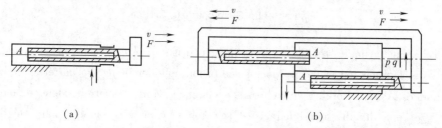

Figure 2-2-4 Plunger Cylinder

The thrust and velocity output by the plunger cylinder can be calculated as follows
$$F = p_A = p \pi d^2 / 4 \quad (2\text{-}20)$$
$$v_i = q/A = 4q / \pi d^2 \quad (2\text{-}21)$$

3 Other hydraulic cylinders

(1) Supercharging cylinder is shown in Figure 2-2-5. Supercharging cylinder, also called supercharger, makes use of the difference between effective areas of the piston and the plunger to get a high pressure in local areas of the hydraulic system. Supercharging cylinders are divided into single-acting type and double-acting type. The working principle of the single-acting supercharging cylinder is shown in Figure 2-2-5(a). With the pressure of fluid flowing into the piston cylinder being recorded as p_1, the piston diameter as D and the plunger diameter as d, the output pressure of the fluid from the plunger cylinder is high and can be calculated as follows:
$$p_2 = p_1 (D/d)^2 = K p_1 \quad (2\text{-}22)$$

Where: $K = (D/d)^2$, which is called supercharging ratio and represents the degree of supercharging.

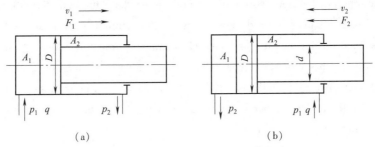

Figure 2-2-2 Single-Rod Piston Cylinder

Due to unequal effective working areas of the two chambers of the hydraulic cylinder, the output thrust and velocity in the two directions are different and can be calculated as follows

$$F_1 = (p_1 A_1 - p_2 A_2) = \pi[(p_1 - p_2)D^2 - p_2 d^2]/4 \tag{2-14}$$

$$F_1 = (p_1 A_1 - p_2 A_2) = \pi[(p_1 - p_2)D^2 - p_2 d^2]/4 \tag{2-15}$$

$$v_1 = q/A_1 = 4q/\pi D^2 \tag{2-16}$$

$$v_2 = q/A_2 = 4q/\pi(D^2 - d^2) \tag{2-17}$$

According to Formulae (4-14) to (4-17), because of $A_1 > A_2$, $F_1 > F_2$ and $v_1 < v_2$ can be inferred. If the ratio of output velocity v_2 to v_1 is called speed ratio and recorded as λ_v, $\lambda_v = v_2/v_1 = 1/[1-(d/D)^2]$. Accordingly, $d = D\sqrt{(\lambda_v - 1)/\lambda_v}$. The value of d can be determined if D and λ_v are known.

Differential cylinder. The single-rod piston cylinder forms "differential connection" if both of its left and right chambers are supplied with high pressure fluid, as shown in Figure 2-2-3. The fluid pressures in left and right chambers of the cylinder with differential connection are the same. However, because the effective area of the left chamber (without rod) is larger than that of the right chamber (with rod), when the piston moves rightward, the fluid (with a flow of q') is discharged from the right chamber into the left chamber, thus increasing the flow $(q+q')$ into the left chamber and accelerating the movement of the piston. Actually, due to the pressure loss in the pipeline of the two chambers with differential connection during piston movement, the fluid pressure in the right chamber differs from (is slightly greater than) that in the left chamber. However, the difference is generally small and can be ignored. The thrust (F_3) and movement velocity (v_3) of the piston with differential connection can be calculated as follows

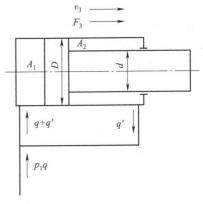

Figure 2-2-3 Differential Cylinder

machine tools in general. If a long stroke of the table is needed, the cylinder with fixed piston rod shown in Figure 2-2-1(b) may be used. For this type, the cylinder block is connected to the table, the piston rod is fixed to the machine tool through a bracket, and the power is transmitted via the cylinder block. As for this mounting type, the moving range of the table is only two times the effective stroke l of the hydraulic cylinder (i. e. $2l$), so the hydraulic cylinder covers a small area. However, its fluid inlet and outlet ports may be set at both ends of the stationary hollow piston rod, but they must be connected with a hose.

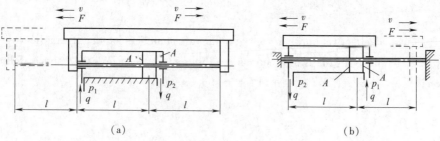

Figure 2-2-1 Double-Rod Piston Cylinder

Due to the normally equal diameter of piston rods at both ends of the double-rod piston cylinder, the effective area of the left chamber is equal to that of the right chamber. When any fluid with the same pressure and the same flow rate is input into the left and right chambers respectively, the thrust and velocity of the hydraulic cylinder in left and right directions are the same. With the piston diameter being recorded as D, the piston rod diameter as d, the fluid chamber inlet and outlet pressures of the hydraulic cylinder as p_1 and p_2, and the input flow rate as q, the thrust F and velocity v of the double-rod piston cylinder can be calculated as follows

$$F = A(p_1 - p_2) = \pi(D^2 - d^2)(p_1 - p_2)/4 \qquad (2\text{-}12)$$

$$v = q/A = 4q/\pi(D^2 - d^2) \qquad (2\text{-}13)$$

Where: A is the effective working area of piston.

In terms of the design of double-rod piston cylinder, one of its piston rods bears tensile stress and the other bears no stress during operation. Therefore, the piston rod of such cylinder may be slender.

(2) Single-rod piston cylinder. As shown in Figure 2-2-2, single-rod piston cylinders are divided into fixed cylinder block type and fixed piston rod type. For such cylinders, only one end of the piston is provided with a piston rod, but their moving range of table is two times the effective stroke of piston.

Continued

Type	Name	Symbol	Description
Double-acting hydraulic cylinder	Hydraulic cylinder with single piston rod		The rod is only provided for one side of the piston. The two-way hydraulic drive is adopted. The thrust and speed in two ways are not equal.
	Hydraulic cylinder with double piston rods		Piston rods are provided for both sides of the piston. The two-way hydraulic drive is adopted. It can realize the reciprocating motions at a constant speed.
	Telescopic hydraulic cylinder		The two-way hydraulic drive is adopted. The cylinder extends from the largest stage to the smallest, and retracts from the smallest stage to the largest.
Combined hydraulic cylinder	Spring resetting hydraulic cylinder		One-way hydraulic drive is adopted. The cylinder is reset by spring force.
	Series hydraulic cylinder		It is used to obtain large thrust when the diameter of the cylinder is limited while the length is not limited.
	Supercharging cylinder (supercharger)		It is driven by the low-pressure chamber A to provide the high-pressure fluid source for chamber B.
	Rack-and-pinion hydraulic cylinder		The piston drives the gear by reciprocating motion to rotate back and forth via the rack and pinion installed together.
Swing hydraulic cylinder	Swing motor		The output shaft directly outputs the torque, and its reciprocating rotation angle is less than 360°. It is also called swing motor.

II Commonly Used Hydraulic Cylinders

1 Piston type hydraulic cylinder

Piston type hydraulic cylinders are divided into double-rod type and single-rod type by application requirements.

(1) The double-rod piston cylinder is shown in Figure 2-2-1. The hydraulic cylinder with piston rods of the same diameter extending at both ends of the piston is called double-rod piston cylinder, which generally consists of cylinder head, piston, cylinder block, piston rod, seal, etc. Double-rod piston cylinders are divided into fixed barrel type and fixed piston rod type by installation methods.

Figure 2-2-1(a) shows a double-rod piston cylinder with fixed barrel. Its fluid inlet and outlet ports are arranged at both ends of the barrel and the piston drives the table to move via the piston rod. When the effective stroke of piston is l, the moving range of the entire table will be $3l$. Because the cylinder covers a large area on the machine tool, it is not applicable to small-size

Task implementation

1 Draw the graphic symbols of hydraulic motors on the equipment.
2 Skillfully regulate the speed of the motor.
3 Discuss in groups and summarize the application scenarios of various hydraulic motors.

Task 2   Knowledge about Hydraulic Cylinder

 Task description

Understand the basic working principles and symbols of common hydraulic cylinders, be acquainted with the types of common hydraulic cylinders, and be familiar with the applications of common hydraulic cylinders in different scenarios.

Relevant knowledge

I Types, Characteristics and Symbols of Hydraulic Cylinders

Hydraulic cylinder, also called fluid cylinder, is an actuator in the hydraulic system and it is used to convert hydraulic energy into mechanical energy.

There are many types of hydraulic cylinders. Refer to Table 2-2-1 for details about their types and symbols.

Table 2-2-1 Types, Characteristics and Symbols of Common Hydraulic Cylinders

Type	Name	Symbol	Description
Single-acting hydraulic cylinder	Plunger type hydraulic cylinder		The plunger only moves in one direction, and returns by its own weight or load.
	Hydraulic cylinder with single piston rod		The piston only moves in one direction, and returns by its own weight or load.
	Hydraulic cylinder with double piston rods		Piston rods are provided for both sides of the piston. The pressure fluid is only supplied to one side. Generally, the piston returns by spring force, gravity or external force.
	Telescopic hydraulic cylinder		It has a long stroke with a short cylinder. Hydraulic fluid is used to extend the cylinder from the largest stage to the smallest, and retract it from the smallest stage to the largest by an external force.

$$i = n_{max}/n_{min} \qquad (2-11)$$

III Working Principle of Hydraulic Motor

The structure of commonly used hydraulic motors is very similar to that of hydraulic pumps in the same category. The following text introduces the working principle by taking vane motor as an example.

Figure 2-1-3 shows the working principle diagram of a vane type hydraulic motor.

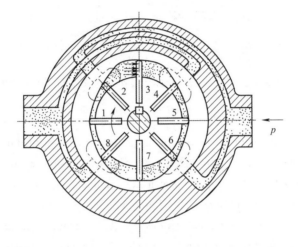

Figure 2-1-3 Working Principle Diagram of Vane Motor

When any fluid with the pressure of p flows between vanes 1 and 3 via the fluid inlet port, both sides of vane 2 is under the effect of hydraulic fluid and no torque is generated. As for vanes 1 and 3, pressure fluid exists on one side and low-pressure fluid exists on the other side. However, because the extending area of vane 3 is larger than that of vane 1, the total hydraulic pressure on vane 3 is greater than that on vane 1 and the differential pressure makes the rotor generates a clockwise torque. Similarly, when any hydraulic fluid flows between vanes 5 and 7, the extending area of vane 7 is larger than that of vane 5 and a clockwise torque is also generated. In this way, the fluid pressure is converted into mechanical energy. This is the working principle of vane motor. The hydraulic motor rotates reversely when the fluid delivery direction is changed.

The vane motor outputs a higher torque in the case of greater difference between inner and outer diameters of the stator, the longer the rotor diameter and the higher the input pressure.

The small volume of vane motor results in its low rotational inertia, quick action and high reversing frequency. However, due to its high leakage amount, vane motor cannot run at a very low speed, so it is generally applied to the scenarios requiring high speed, small torque and quick action.

movement of the spring causes the compression of the spring and the increase of thrust, the object moves and jumps again. The whole process forms a state of intermittent movement and stop shown in Figure 2-1-2(b). For a hydraulic motor, this is creeping.

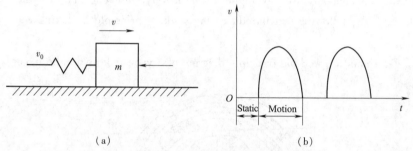

Figure 2-1-2 Physical Model for Creeping of Hydraulic Motor

(2) The leakage amount is unstable.

The leakage amount of hydraulic motor fluctuates periodically with the change of phase angle during the rotation of the rotor, so it is not the same at each moment. As the flow to the motor running at a low speed is low, the proportion of leakage amount becomes relatively high. Unstable leakage amount will obviously influence the flow rate of the running motor, resulting in the fluctuation of speed. When the motor runs at a low speed, its rotating part and the load driven by it show less inertia and obvious influence mentioned before, thus causing creeping. In practical work, the minimum steady speed is expected to be "the smaller, the better" in general.

7 Maximum operating speed

The maximum operating speed of hydraulic motor is mainly restricted by its service life and mechanical efficiency. With the increase of speed, the wear of each moving pair is intensified, thus shortening the service life. High speed means that the hydraulic motor needs a high input flow, so the flow rate of each flow passage part increases correspondingly and the pressure loss also rises, thus reducing the mechanical efficiency.

For some hydraulic motors, the increase of speed is restricted by back pressure. For example, as for a crankshaft connecting rod type hydraulic motor, if the speed is increased, the fluid return back pressure must be increased significantly to ensure no bumping of the connecting rod against the crankshaft surface. With the increase of speed, the back pressure for the fluid return chamber also increases. However, excessive increase of back pressure results in obvious reduction in the efficiency of the hydraulic motor. To avoid too low efficiency of the hydraulic motor, the motor's speed should not be too high.

8 Speed regulation range

The speed regulation range of hydraulic motor is expressed by the ratio of the maximum operating speed to the minimum steady speed, namely

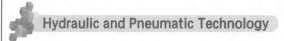

Table 2-1-2 shows that, in terms of starting performance, the multi-acting inner curve motor is the best, the axial plunger motor, crankshaft connecting rod type motor and hydrostatic balance motor are good, the vane motor is poor, and the gear motor is the worst.

5 Speed of hydraulic motor

The speed of hydraulic motor depends on the flow rate of the fluid and the displacement V of the hydraulic motor itself, which can be calculated with the following formula:

$$n_t = q_i / V \qquad (2\text{-}9)$$

Where: n_t is the theoretical speed (in r/min).

Because some fluid is lost due to the leakage inside the hydraulic motor, not all the fluid entering the hydraulic motor drives it to work. As a result, the hydraulic motor's actual speed is less than its theoretical speed.

$$n = n_t \eta_v \qquad (2\text{-}10)$$

Where: n is the actual speed of hydraulic motor (in r/min); η_v is the volumetric efficiency of hydraulic motor (in %).

6 Minimum steady speed

Minimum steady speed refers to the minimum speed of the hydraulic motor under the rated load without creeping. Creeping means that when the hydraulic motor runs at a very low speed during operation, it always fails to maintain at a constant speed and enters an unstable state with intermittent movement and stop.

The reasons why the hydraulic motor creeps at a low speed are as follows:

(1) The friction force is unstable.

On the one hand, the friction force usually increases with the increment of speed. However, for a static motor or a motor running at a low speed, its internal frictional resistance does not increase but decreases with the increment of speed, forming the so-called "negative characteristic" resistance. On the other hand, both the hydraulic motor and the load are driven by the rising pressure after the hydraulic fluid is compressed. Therefore, the working process of the hydraulic motor running at a low speed can be expressed by the physical model shown in Figure 2-1-2(a): one end of the spring (equivalent to the non-compressible working medium under high pressure) is driven at a constant speed v_0, so that the object with a mass of m (equivalent to the mass and rotational inertia of both the motor and the load) can move by overcoming the "negative characteristic" frictional resistance. When the object keeps stationary or moves at a low speed, the resistance will be high and the spring constantly compressed. At this moment, it is necessary to increase the thrust. The object starts to move only when the spring is compressed until its thrust is greater than the static friction force. Once the object starts to move, if the resistance suddenly decreases, the object will suddenly accelerate. As a result, the compression of the spring will decrease, the thrust will drop and the object will move forward by inertia and then stop. When the

$$T_t = \Delta P \cdot V/2\pi \tag{2-6}$$

Where: ΔP is the differential pressure at fluid inlet and outlet ports of the motor.

3 Mechanical efficiency of hydraulic motor

Due to various types of inevitable friction inside the hydraulic motor, the actual output torque T is always less than the theoretical torque T_t, namely

$$T = T_t \eta_m \tag{2-7}$$

Where: η_m is the mechanical efficiency of hydraulic motor (in %).

4 Starting mechanical efficiency (η_m) of hydraulic motor

The starting mechanical efficiency of hydraulic motor refers to the ratio of the hydraulic motor's actual output torque T_0 to its theoretical torque T_t at the same working differential pressure when the motor starts from a static state, namely

$$\eta_m = T/T_t \tag{2-8}$$

The starting mechanical efficiency of the hydraulic motor is also an indicator of its starting performance. Under the same pressure, when the hydraulic motor starts from a static state, its output torque is greater than that during its operation. This makes it difficult to start the hydraulic motor with load, so the starting performance is very important for the hydraulic motor and the starting mechanical efficiency can reflect its starting performance.

In practical work, good starting performance is always desired, i.e. higher starting torque and starting mechanical efficiency are expected. The approximate values of starting mechanical efficiency η_{m0} of hydraulic motors with different structures are listed in Table 2-1-2.

Table 2-1-2 Starting Mechanical Efficiency of Hydraulic Motor

Structure of hydraulic motor		Starting mechanical efficiency η_{m0}/%
Gear motor	Old structure	0.60-0.80
	New structure	0.85-0.88
Vane motor	High-speed type with a small torque	0.75-0.85
Axial plunger motor	Piston shoe type	0.80-0.90
	Non-piston shoe type	0.82-0.92
Crankshaft connecting rod type motor	Old structure	0.80-0.85
	New structure	0.83-0.90
Hydrostatic balance motor	Old structure	0.80-0.85
	New structure	0.83-0.90
Multi-acting inner curve motor	The tangential force is transferred by the sliding friction pair on the cross beam	0.90-0.94
	A rolling pair exists at the part transferring the tangential force	0.95-0.98

II Performance Parameters of Hydraulic Motor

There are many performance parameters of hydraulic motor. The main performance parameters are as follows:

1 Displacement, flow rate and volumetric efficiency

Generally, the volume of entering fluid calculated as per geometric dimensions for each rotation of the motor shaft is called the displacement V of the motor, also called theoretical displacement or geometric displacement, i.e., the displacement without considering the leakage loss.

The displacement of a hydraulic motor represents the size of its working chamber. During operation, the output torque is decided by load torque. Therefore, displacement is an important parameter. However, to drive the same load, the motor with a larger working chamber bears a pressure lower than that with a smaller working chamber. For this reason, the size of working chamber is the main indicator of the working capacity of the hydraulic motor. That is to say, displacement is an important sign of the working capacity of the hydraulic motor.

It can be known from the working principle of hydraulic power elements that the relation among the speed n, theoretical flow rate q_i and displacement V of the motor is as follows

$$q_i = nV \tag{2-1}$$

Where: q_i is the theoretical flow rate (in m³/s); n is speed (in r/min); V is displacement (in m³/s).

To meet the speed requirement, the actual input flow rate q of the motor is usually higher than the theoretical flow rate

$$q = q_i + \Delta q \tag{2-2}$$

Where: Δq is the leakage flow rate.

$$\eta_v = q_i/q = 1/(1+\Delta q/q_i) \tag{2-3}$$

Accordingly, the actual flow rate is

$$q = q_i/\eta_v \tag{2-4}$$

2 Theoretical torque output by hydraulic motor

Based on the displacement, it is able to calculate the torque that can be output by the hydraulic motor with a given pressure or calculate the working pressure of the hydraulic motor with a given load torque. With the differential pressure at fluid inlet and outlet ports of the hydraulic motor being recorded as ΔP, the input flow rate of the hydraulic motor as q, the theoretical torque output by hydraulic motor as T_t and the angular velocity as ω, if no loss is counted, the hydraulic power input to the hydraulic motor should be completely converted into the mechanical power output by it, namely

$$\Delta P \cdot q = T_t \omega \tag{2-5}$$

And because $\omega = 2\pi n$, the theoretical torque of the hydraulic motor is

adopted. The reason is that it is hard to form lubricating film if any dynamic pressure bearing is used at a very low motor speed.

(4) Relying on the high-speed rotation of its vane and rotor, the vane pump generates a centrifugal force which can keep the vane closely touch with the inner surface of the stator, thus sealing the fluid and forming a working volume. If the vane pump is used as a motor, a spring must be installed at the vane root of the hydraulic motor to ensure that the vane closely touches with the inner surface of the stator, so that the motor can start normally.

(5) Structurally, it is required to ensure the self-priming capacity of a hydraulic pump, but for a hydraulic motor, no such requirement exists.

(6) Hydraulic motor must have a large starting torque. The so-called starting torque refers to the torque that can be output by the motor shaft when the motor starts from a static state. This torque is usually greater than the torque of the motor running at the same working differential pressure. Therefore, to make the starting torque as close as possible to the torque in the working state, both the torque fluctuation and internal friction of the motor should be at a low level.

Due to their different characteristics mentioned above, many types of hydraulic pumps and hydraulic motors cannot be used interchangeably.

Hydraulic motors can be divided into low-speed and high-speed types by rated speed. The ones with a rated speed lower than 500 r/min are low-speed hydraulic motors, while those with a rated speed higher than 500 r/min are high-speed hydraulic motors.

Hydraulic motors can also be divided into gear, vane, plunger and other types by structure.

Graphical symbols of hydraulic motors see Table 2-1-1.

Table 2-1-1 Symbols of Hydraulic Motors

Name	Symbol	Name	Symbol
Fixed-displacement hydraulic pump/motor		Two-way fixed-displacement motor	
Variable displacement hydraulic pump/motor		One-way variable displacement motor	
One-way fixed-displacement motor		Two-way variable displacement motor	

Part 2
Hydraulic Actuators

Task 1 Knowledge about Hydraulic Motor

Task description

Understand the basic working principles and symbols of common hydraulic motors, and get familiar with the types of hydraulic motors and the applications of common hydraulic motors in different scenarios.

Relevant knowledge

I Concept and Symbol of Hydraulic Motor

Hydraulic motor is a device that converts the pressure energy of fluid into mechanical energy, see Figure 2-1-1. In structure, hydraulic pump and hydraulic motor in the same category are very similar, but they differ a little in structure due to their different working conditions. In principle, hydraulic motor can also be used as hydraulic pump and vice versa. The main differences between them are as follows:

(1) Generally, hydraulic pump rotates in only one direction, but hydraulic motor performs forward and reverse rotation. Therefore, they are symmetrical in internal structure.

(2) To reduce radial force and fluid suction resistance, the size of the fluid suction port of a hydraulic pump is usually larger than that of its fluid outlet, while the pressure in the low pressure chamber of a hydraulic motor is slightly higher than the atmospheric pressure.

Figure 2-1-1 Real Product Picture of Hydraulic Motor

(3) Hydraulic motor is required to work normally in a wide speed range, so hydrodynamic or hydrostatic bearings should be

2 Grouping and discussion: groups of 3-5. Summarize the differences among commonly used hydraulic pumps such as vane pumps, gear pumps and plunger pumps.

3 Find out the model of the hydraulic pump used in a common hydraulic equipment station, such as the hydraulic clamping system in the numerically controlled lathe. Write the selection basis.

Thinking and Exercise

1 Find out the model of the hydraulic pump used in a hydraulic station, and write the parameters such as pressure and flow rate according to the nameplate.

2 What performance needs to be considered for selecting a hydraulic pump?

3 Summarize some basic characteristics and common application scenarios of hydraulic pumps that we have learned.

Relevant knowledge

The hydraulic pump is a fluid power element that provides a certain flow rate and pressure. It is an indispensable core element of every hydraulic system. Selecting a suitable hydraulic pump is very important to reduce energy consumption and noise of the hydraulic system, improve system efficiency and working performance, and ensure reliable operation.

Principles of selecting a hydraulic pump: Determine the type of hydraulic pump based on the working conditions of the main machine, power and working performance required by the system, and then determine the specifications and model according to the pressure and flow rate required by the system and considering the price and maintenance. Table 1-5-1 lists the main performance and application scenarios of hydraulic pumps commonly used in hydraulic systems.

Table 1-5-1 Performance Comparison of Hydraulic Pumps Commonly Used in Hydraulic Systems

Performance	Externally-engaged gear pump	Double-acting vane pump	Pressure-limited variable vane pump	Radial plunger pump	Axial plunger pump	Screw pump
Output pressure	Low pressure	Medium pressure	Medium pressure	High pressure	High pressure	Low pressure
Flow adjustment	Non-adjustable	Non-adjustable	Adjustable	Adjustable	Adjustable	Non-adjustable
Efficiency	Low	Relatively high	Relatively high	High	High	Relatively high
Output flow fluctuation	Very large	Very small	General	General	General	Minimum
Self-priming performance	Good	Relatively poor	Relatively poor	Poor	Poor	Good
Sensitivity to fluid pollution	Insensitive	Relatively sensitive	Relatively sensitive	Very sensitive	Very sensitive	Insensitive
Noise	Large	Small	Relatively large	Large	Large	Minimum

In general, as each type of hydraulic pump has particular characteristics and the structures, functions and rotation modes of different types are different, a suitable hydraulic pump should be selected according to the specific application scenario. Double-acting vane pumps or pressure-limited variable vane pumps are always adopted for the machine tool hydraulic systems. For the road construction machinery, harbor machinery and small engineering machinery, gear pumps that are insensitive to fluid pollution are often used. Plunger pumps are often selected for those requiring heavy load and high power.

Task implementation

1 Look up the selection principles of the hydraulic pump online, and summarize them.

Task implementation

1 After learning, select a plunger pump suitable for the double housing planer, and determine its power, allowable maximum pressure, flow rate, and other parameters.

2 Grouping and discussion: groups of 3-5. List the differences among the vane pump, gear pump and plunger pump.

Task 5   Selection of Hydraulic Pump

Task description

The hydraulic station of a hydraulic equipment platform, as shown in Figure 1-5-1 below, is provided with two hydraulic pumps. One is the variable vane pump shown in Figure 1-5-2(a), and the other one is the plunger pump shown in Figure 1-5-2(b). Please list the reasons why these two pumps are selected by looking up relevant materials such as equipment specifications or according to the knowledge learned.

Figure 1-5-1   Hydraulic Station of a Hydraulic Equipment Platform

(a) Variable vane pump

(b) Plunger pump

Figure 1-5-2   Hydraulic Pumps

2 Structural characteristics of axial plunger pump

Figure 1-4-4 below shows a structural diagram of an axial plunger pump. The transmission shaft 6 drives the cylinder body 3 to rotate via the splines. Seven plungers 7 are evenly installed on the cylinder block. The head of the plunger is provided with a sliding shoe 9. The sliding shoe is connected with the plunger by a spherical hinge. It can rotate freely. The shoe is pressed against the swashplate 11. In this way, when thecylinder rotates, the plunger can reciprocate in the cylinder to complete the fluid suction and pressing process. The fluid distribution pan 4 is fixed on the pump body and connected with the fluid suction and pressing ports of the pump. Furthermore, there is a fluid chamber at the part where the sliding shoe contacts with the swashplate. The pressure fluid enters the fluid chamber via the small hole in the middle of the plunger, forming a fluid film between the sliding shoe and swashplate. The film provides static pressure support to reduce wear.

The variable mechanism of the axial plunger pump is manual. To change the output flow of the pump, the hand wheel 15 needs to be turned to adjust the inclined angle of the swashplate by the lead screw nut pair.

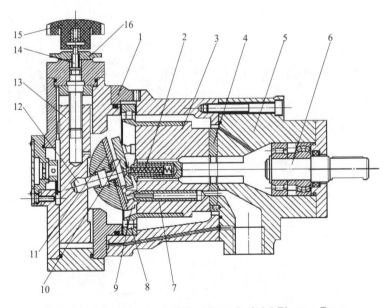

Figure 1-4-4 Structure of Straight-axis Axial Plunger Pump

1-Pump body;2-Spring;3-Cylinder body;4-Fluid distribution pan;5-Front pump body;
6-Transmission shaft;7-Plunger;8-Bearing;9-Sliding shoe;10-Slipper retainer plate;
11-Swashplate;12-Shaft pin;13-Variable piston;14-Lead screw;15-Hand wheel;16-Nut

in Figure 1-4-3, when the cylinder body angle is in the range of π-2π, the plunger extends outward to increase the enclosed working volume of the cylinder hole at the bottom of the plunger, so that the fluid is sucked via the fluid suction window of the fluid distribution pan; in the range of 0-π, the plunger is pushed into the cylinder block by the swashplate to reduce the cylinder hole volume so that the fluid is pressed via the fluid pressing window of the pan. As the cylinder body rotates for one cycle, the fluid suction and pressing actions are completed once by each plunger. The axial plunger pump features a compact structure, high working pressure and easy variable realization.

As shown in (b) of Figure 1-4-3, the cylinder axis of the inclined-axis axial plunger pump is inclined at an angle with the transmission shaft axis. The end of the transmission shaft is connected with each plunger in the cylinder by a universal hinge and connecting rod. When the transmission shaft rotates, the hinge and rod drive the plunger and cylinder to rotate together (the plunger is forced to reciprocate in the cylinder), so as to complete the fluid suction and pressing via the fluid distribution pan. This kind of pump has the advantages of large variable range and high pump intensity. However, compared with the straight-axis type mentioned above, it has a more complex structure, larger overall dimensions, and higher weight.

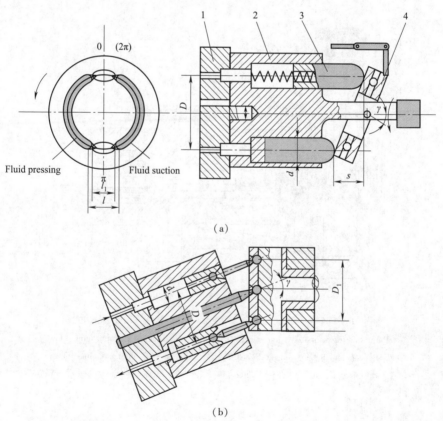

Figure 1-4-3 Working Principle of Axial Plunger Pump
1-Cylinder body; 2-Fluid distribution pan; 3-Plunger; 4-Swashplate

working space of the pump. Obviously, when the rotor rotates in the direction shown in the figure, the working space located in the upper part is in a fluid suction state (the fluid in the tank enters the chamber b via the hole a of the fluid distribution shaft); the working space located in the lower part is in a fluid pressing state (the fluid in the chamber c is output via the hole d of the fluid distribution shaft). The output flow and fluid suction and pressing directions of the pump can be changed by adjustment of the size and direction of eccentricity e between the stator and rotor.

Figure 1-4-1 Real Product Picture of Plunger Pump

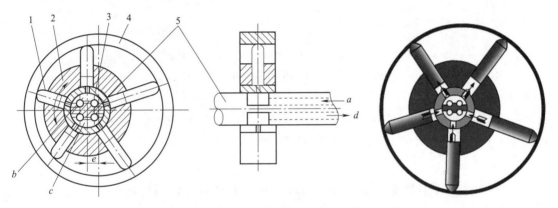

Figure 1-4-2 Working Principle of Radial Plunger Pump
1-Plunger; 2-rotor; 3-Bushing; 4-Stator; 5-Fluid distribution shaft

II Axial Plunger Pump

1 Working Principle of Axial Plunger Pump

The axial plunger pump is a kind of pump that has various plungers located on the circumference of a common cylinder with the centerline of plungers parallel to that of the cylinder. The axial plunger pumps are divided into the straight-axis type (swashplate) and inclined-axis (swing cylinder). The working principle of the straight-axis axial plunger pump is shown in Figure 1-4-3. The pump body is composed of the cylinder block 1, fluid distribution pan 2, plunger 3 and swashplate 4. Plungers are evenly distributed in the cylinder body along the circumference. The swashplate axis is inclined at an angle with the cylinder block axis. The plunger is pressed against the swashplate (spring in the figure) by a mechanical device or under the action of the low-pressure fluid. The fluid distribution pan 2 and the swashplate 4 are fixed. When the prime mover drives the cylinder block to rotate via the transmission shaft, the plunger is forced to reciprocate in the cylinder block under the action of the swashplate, and the fluid suction and pressing are completed via the fluid distribution window of the pan. Inthe rotating direction as shown

used double-acting vane pump is 2.5-300 mL/r, and that of a variable vane pump is 6-120 mL/r.

(3) Speed: The maximum speed of a small displacement double-acting vane pump can reach 8,000-10,000 r/min, and that of a general displacement vane pump 1,500-2,000 r/min. For a commonly used variable displacement vane pump, the maximum speed is about 3,000 r/min, the minimum speed is 600-900 r/min.

(4) Efficiency: For a double-acting vane pump, the volumetric efficiency is relatively high (up to 93%-95%), but the mechanical efficiency is low. Its total efficiency is similar to that of a gear pump.

(5) Service life: The service life of a vane pump is longer than that of a gear pump. The service life of a high-pressure vane pump can reach more than 5,000 h.

Task implementation

1 Select a vane pump suitable for the hydraulic equipment platform after learning.

2 Grouping and discussion: groups of 3-5. List the differences between the vane pump and gear pump.

3 Look up the materials regarding other commonly used vane pumps (such as limited vane pump), and analyze the working principle of the limited vane pump.

Task 4   Knowledge about Plunger Pump

Task description

Given that a certain type of double housing planer needs to be equipped with a hydraulic pump that features high pressure, large flow, high power, compact structure, taking up less space, high efficiency and convenient flow adjustment, what kind of hydraulic pump should be equipped? Analyze the structure and characteristics of the selected pump.

Relevant knowledge

I   Radial Plunger Pump

The physical plunger pump is shown in Figure 1-4-1, and the working principle of the radial plunger pump in Figure 1-4-2. The plunger pump is composed of a plunger, rotor (cylinder body), bushing, stator and fluid distribution shaft. There is an eccentricity e between the stator and the rotor. The bushing is fixed in the rotor hole. It rotates with the rotor, while the fluid distribution shaft is fixed and stationary. The plunger moves in the radial hole of the rotor to form the enclosed

two pumps may be equal or not equal.

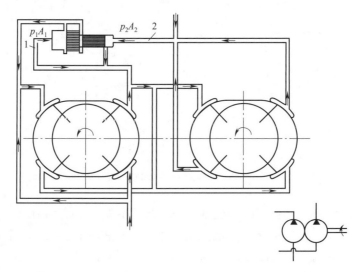

Figure 1-3-7　Working Principle of Two-stage Vane Pump
1,2-Pipeline

Double vane pumps are often used in special machining machine tools with fast feeding and working feeding requirements. In this case, the double pump is composed of a small flow pump and a large flow pump. During fast feeding, both pumps supply fluid at the same time (at this time, the pressure is relatively low); during working feeding, the small flow pump supplies fluid (at this time, the pressure is relatively high), and the large flow pump is subject to unloading on the fluid circuit system. Compared with high-pressure large flow pumps, double vane pumps have the advantages of saving energy and reducing fluid heat. Such pumps are also commonly used in the machine tool hydraulic systems, which require two independent fluid circuits that do not affect each other.

3　Main performance of vane pump

Compared with the gear pump, the vane pump has a more complex structure, but it features higher working pressure, small flow fluctuation, stable operation, low noise and longer service life. Therefore, it is widely used for special machine tools, automatic lines, and other medium- and low-pressure hydraulic systems in the mechanical manufacturing process. However, its structure is complex, and the fluid suction performance is not very good. Moreover, it is sensitive to fluid pollution. The main performance parameters of the vane pump are as follows:

(1) Pressure: The rated pressure of a medium- and low-pressure vane pump is generally 6.3 MPa, the maximum working pressure of a double-acting high-pressure vane pump can reach 28-30 MPa, and the pressure of a variable vane pump is not greater than 17.5 MPa in general.

(2) Displacement: The displacement of a vane pump is 0.5-4,200 mL/r, that of a commonly

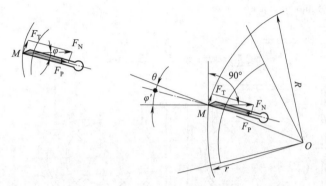

Figure 1-3-6 Inclined Angle of Double-acting Vane Pump

3 Measures to increase the pressure of double-acting vane pump

The hydraulic force on the vane top and bottom in the fluid suction area is unbalanced because pressure fluid exists at the vane bottom of the double-acting vane pump generally. The vane top abuts against the inner surface of the stator fluid suction area with a very high compression force, which aggravates the wear and affects the service life of the vane pump, especially when the working pressure is high. Accordingly, the unbalanced pressure at both ends of vanes in the fluid suction area limits the increase of the working pressure of the double-acting vane pump. Therefore, measures must be taken for the structure of the high-pressure vane pump.

Commonly used measures to reduce the force of the vane pressing against the stator:

(1) Reduce the fluid pressure acting on the vane bottom.

(2) Reduce the area under the action of pressure fluid at the vane bottom.

(3) Ensure that the hydraulic force at the vane top and bottom is balanced.

III Two-stage Vane Pump and Double Vane Pump

1 Two-stage vane pump

A two-stage vane pump can be used instead of a high-pressure vane pump to obtain relatively high working pressure. The two-stage vane pump is composed of two single-stage vane pumps with the pressure circulated, which are installed in a pump body and connected in series on the fluid circuit. The working principle of the two-stage vane pump is shown in Figure 1-3-7. The rotors of two single-stage vane pumps are installed on the same transmission shaft. As the transmission shaft rotates, the two rotors are driven to rotate together, which can generate double working pressure.

2 Double vane pump

The double vane pump is composed of two single-stage vane pumps, which are installed in a pump body and connected in parallel on the fluid circuit. The rotors of the two vane pumps are driven to rotate by the same transmission shaft. They have independent fluid outlets. The flow of

area to a high-pressure area), the fluid in the fluid pressing chamber will inevitably flow backward, resulting in the output flow and pressure fluctuation. If, on the fluid distribution pan, the vanes enter the triangular groove on one side of the fluid pressure window from the fluid sealing area, the low-pressure fluid will gradually flow to the fluid pressing window to increase the pressure, thereby reducing the flow and pressure fluctuation of the pump. The dimensions of triangular grooves are usually determined by experiments.

(2) Stator curve.

The transition curve of the stator is shown in Figure 1-3-5. It consists of four-section arcs and four-section transition curves. The ideal transition curves should ensure that the radial speed and acceleration change uniformly when the vanes slide in the rotor slot, and ensure that the impact of the vanes on the stator surface is as small as possible.

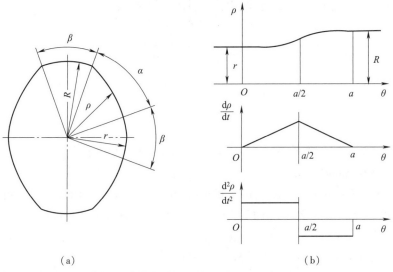

Figure 1-3-5 Transition Curve of Stator

(3) Inclined angle of vane.

The vanes should be arranged properly in the rotor to ensure easy sliding of vanes and less wear. The inclined angle of double-acting vane pump is shown in Figure 1-3-6 below. In the working process, the vanes tightly contact with the stator under the action of centrifugal force and pressure fluid at the vane bottom. The counter-acting force N acting on the cane top by the inner surface of the stator can be decomposed into two forces, namely, the force T perpendicular to the vane and the force P along the vane slot direction. The force T is easy to break the vane. As a result, the rotor slot is usually inclined at an angle of α along the rotation direction to reduce the value of the force T. Based on theoretical analysis and experimental verification, α is generally taken as $10°$-$14°$.

pressing chambers, and their central angles are symmetrical. Consequently, the fluid pressure acting on the rotor is balanced. To balance the radial force completely, the number of enclosed spaces (i.e. number of vanes) should be even.

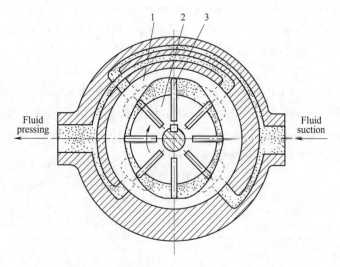

Figure 1-3-3 Working Principle of Double-acting Vane Pump
1-Stator; 2-Rotor; 3-Vanes

2 Structural characteristics of double-acting vane pump

(1) Fluid distribution pan.

The fluid distribution pan of the double-acting vane pump (namely, fluid distribution mechanism of the pump) is shown in Figure 1-3-4. To ensure that the fluid suction and pressing windows of the fluid distribution pan can be separated during operation, the included angle α of the fluid sealing area on the pan must be greater than or equal to the included angle between two adjacent vanes. In addition, the included angle β of the stator arc must $\beta \geq \alpha$ to avoid fluid pocketing and cavitation. As shown in Figure 1-3-4, a triangular groove is provided on the fluid pressing window of the pan. It is mainly used to reduce the flow and pressure fluctuation of the pump. The pressure of fluid between two adjacent vanes in the fluid sealing area is basically the same as that in the fluid suction area. If the fluid reaches the fluid pressure window from the fluid sealing area (equivalent to the sudden connection of a low-pressure

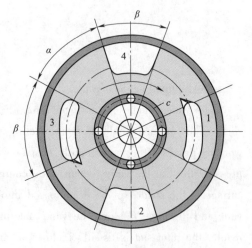

Figure 1-3-4 Fluid Distribution Pan of Double-acting Vane Pump

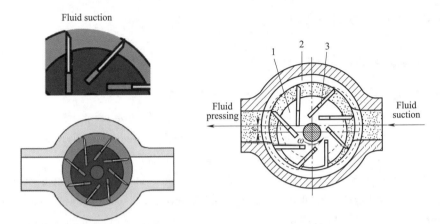

Figure 1-3-2 Working Principle of Single-acting Vane Pump
1-Rotor;2-Stator;3-Vanes

(3) As the rotor is subject to the unbalanced radial hydraulic force, this kind of pump is generally unsuitable for high pressure.

(4) To facilitate the outward extension of vanes under the action of inertial force, an inclination angle (which is called caster angle) opposite to the rotation direction should be formed. Generally, the angle is 24°.

II Double-acting Vane Pump

1 Working principle of double-acting vane pump

The working principle of the double-acting vane pump is shown in Figure 1-3-3. The pump is also composed of stator 1, rotor 2, vanes 3 and fluid distribution pan. The rotor center coincides with the stator center, and the inner surface of the stator is approximately elliptical. This ellipse consists of two-section long radii R, two-section short radii r and four-section transition curves. When the rotor rotates, the vanes, under the action of centrifugal force, move radially in the rotor slot and press against the inner surface of the stator, thus forming several enclosed working spaces among the vanes, the inner surface of the stator, the outer surface of the rotor and the fluid distribution pans on both sides. When the rotor rotates in the direction shown in the figure, the enclosed space on the small arc moves to the big arc viathe transition curve, and the vanes extend outwards, increasing the volume of the enclosed space so that the fluid is sucked. When the space moves from the big arc to the small arc via the transition curve, the vanes are gradually pressed into the groove by the inner wall of the stator, decreasing the volume of the enclosed space so that the fluid is forced out from the fluid pressing port. Obviously, as the inner rotor rotates for one cycle, the fluid suction and pressing actions are completed twice in each enclosed space, so the pump is called double-acting vane pump. The vane pump has two fluid suction chambers and two fluid

Relevant knowledge

I Single-acting Vane Pump

1 Working principle of single-acting vane pump

The single-acting vane pump is shown in Figure 1-3-1 and its working principle in Figure 1-3-2. The single-acting vane pump is composed of rotor 1, stator 2, vanes 3 and end cover. The stator has a cylindrical inner surface. There is eccentricity between the stator and the rotor. Installed in the rotor slot, the vanes can slide in the slot. When the rotor rotates, the vanes abut on the inner wall of the stator due to centrifugal force, thus forming several enclosed working spaces among the stator, rotor, vanes and fluid distribution pans on both sides.

When the rotor rotates in the direction shown in the figure, the vanes in the right part of the figure gradually extend out, and the working space between the vanes gradually increases to suck the fluid via the fluid suction port. This space is the fluid suction chamber. In the left part of the figure, the vanes are gradually pressed into the slot by the inner wall of the stator, and the working space is gradually reduced to force the fluid out via the fluid pressing port. This space is the fluid pressing chamber. Between the fluid suction chamber and the fluid pressing chamber, there is a fluid sealing area that separates the two chambers. For this kind of vane pump, when the rotor rotates for one cycle, the fluid suction and pressing actions are completed once in each working space. Therefore, it is called single-acting vane pump. As the rotor keeps rotating, the pump keeps sucking and discharging fluid.

Figure 1-3-1 Real Product Picture of Vane Pump

2 Structural characteristics of single-acting vane pump

(1) The flow rate can be changed by adjusting the eccentricity between the stator and the rotor.

(2) Under the action of pressure fluid, the vane top in the fluid pressing chamber pushes the vanes into the rotor slot. To make the vane top reliably contact with the inner surface of the stator, the vane bottom on one side of the fluid pressing chamber should be connected to the fluid pressing chamber via a special groove. The vane bottom on one side of the fluid suction chamber should be connected to the fluid suction chamber. The vanes here are pushed against the inner surface of the stator only by centrifugal force.

the inner rotor tooth crest A_1 and the outer rotor tooth recess A_2 gradually increases as the rotor rotates, forming a partial vacuum. The fluid is sucked into the enclosed chamber from the fluid distribution window b. The enclosed volume is the maximum when the fluid reaches A'_1 and A'_2, which marks that the fluid suction is completed. When the rotor continues to rotate, the enclosed space filled with fluid gradually decreases, so that the fluid is squeezed and discharged via the other fluid distribution window a. When another inner rotor tooth is fully engaged with the outer rotor tooth recess A_2, the fluid pressing is completed. As the inner rotor rotates for one cycle, the fluid suction and pressing actions are completed once in each enclosed space formed by A_1 and A_2. When the inner rotor rotates continuously, the fluid suction and discharge of the hydraulic pump are completed. Since the tooth profile of the outer rotor is the circular arc that of the inner rotor is the equidistant curve of curtate hypocycloid, the internally-engaged gear is also called internal gearing cycloidal gear pump and rotor pump.

The internally-engaged gear has many advantages, such as compact structure, small volume, less parts, high speed up to 10,000 r/min, stable movement, low noise and high volumetric efficiency. Its disadvantages include large flow fluctuation and complicated manufacturing process of the rotor. At present, the powder metallurgy process is adopted. With the development of industrial technology, the application of cycloidal gear pumps will become broader. The internally-engaged gear may be used as a hydraulic motor as it can rotate in forward and reverse directions.

Task implementation

1 Explain the structure and composition of the gear pump.

2 Grouping and discussion: groups of 3-5. Each student should explain the working principle of the gear pump briefly.

3 Write the advantages, disadvantages and application scenarios of gear pumps.

Task 3 Knowledge about Vane Pump

Task description

The hydraulic equipment should be equipped with a hydraulic pump, featuring compact structure, uniform flow, low noise and stable operation. Generally, the hydraulic pump should be of medium- and low-pressure. The working environment should be relatively clean. Look up the relevant materials to select a hydraulic pump suitable for this equipment, and analyze the working principle, structural characteristics, advantages and disadvantages of the selected pump.

3 Unbalanced radial force

When the gear pump is working, the gear and bearing will withstand the radial hydraulic pressure. As shown in Figure 1-2-2, the fluid suction chamber is on the right side of the pump, and the fluid pressing chamber is on the left side. In the fluid pressing chamber, the gear is subject to hydraulic pressure, and the leaked fluid along the addendum has different pressure (that is, the unbalanced radial force acting on the gear and bearing). The higher the hydraulic pressure is, the greater the unbalanced force is. This accelerates the bearing wear, shortens the service life of the bearing, and even results in the shaft deformation and the friction between the addendum and the inner wall of the pump body. To eliminate the unbalanced radial force, pressure balancing grooves may be provided on some gear pumps, but this will increase the leakage and reduce the volumetric efficiency. For CB-B gear pumps, the fluid pressing chamber is reduced to decrease the action area of hydraulic pressure on the addendum, thereby reducing the unbalanced radial force. Therefore, the diameter of the fluid pressing port is less than that of the fluid suction port. In addition, the radial clearance between the pump body and addendum may be increased to reduce the radial unbalance force.

III Internally-engaged gear

The internally-engaged gear realizes the fluid suction and pressing by the change in the enclosed volume between gears. The real product and working principle of the internally-engaged gear are shown in Figure 1-2-5.

 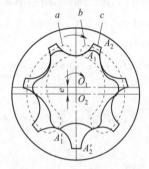

Figure 1-2-5 Real Product and Working Principle of Internally-engaged Gear

It consists of a fluid distribution pan (front and rear covers), an outer rotor (driven gear) and an inner rotor (driving gear) eccentrically arranged in the pump body. The inner rotor has one less tooth than the outer rotor. As shown in Figure 1-2-5, the inner rotor has six teeth and the outer rotor has seven teeth. As the inner and outer rotors are engaged with multiple teeth, several enclosed spaces are formed. When the inner rotor rotates around the center O_1, it drivesthe outer rotor to rotate in the same direction around its center O_2. At this time, the enclosed space formed between

II Structural Characteristics of Gear Pump

1 Leakage

The leakage here refers to the internal leakage of the hydraulic pump (that is, some hydraulic fluid flows back to the fluid suction chamber from the fluid pressing chamber so that it is not delivered to the system). Leakage reduces the volumetric efficiency of the hydraulic pump. For the externally engaged gear pump, leakage may occur between the gear end face and end cover, between the excircle of the gear and the inner hole of the housing, or at the engaging of two gear surfaces. The axial clearance between the gear end face and end cover has the greatest influence on leakage. For this clearance part, the leakage accounts for about 75%-80% of the total leakage due to the short leakage path and large leakage area. Excessively large axial clearance will lead to large leakage and reduce the volumetric efficiency; However, if the clearance is too small, the mechanical friction loss between the gear end face and the end cover will increase, reducing the mechanical efficiency of the pump. Consequently, the axial clearance of the pump must be strictly controlled in design and manufacturing processes.

2 Fluid pocketing of gear pump

To ensure that the gears engage smoothly, based on the engaging principle, the overlap coefficient of gears should be greater than 1 (that is, two pairs of gear teeth can engage at the same time). For this reason, some fluid is pocketed in the enclosed chamber formed by the two pairs of gear teeth, as shown in Figure 1-2-4. The chamber volume first decreases gradually as the gear rotates and then increases gradually. When the volume decreases, the pocketed fluid will generate high pressure due to squeezing, and flow out via the clearance, causing the fluid to heat up and meanwhile the bearing to withstand the unbalanced load. When the volume increases, a partial vacuum will be formed, so that the gas dissolved in the fluid is separated to result in cavitation. This is the fluid pocketing phenomenon of the hydraulic pump. The changein the volume is shown in Figure 1-2-4. In case of fluid pocketing, the gear pump will produce strong noise and cavitation, affecting the working stability and shortening the service life. To eliminate this problem, two unloading grooves for pocketed fluid are provided on the pump cover of the gear pump by milling.

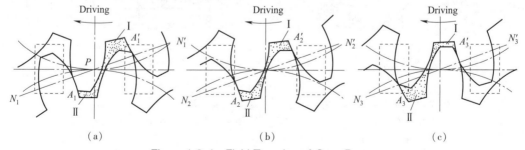

Figure 1-2-4 Fluid Trapping of Gear Pump

addendum line velocity is opposite to the fluid leakage direction. The key point to be considered here is that the clearance should ensure that the addendum does not collide with the inner wall of the pump body when the gear is subject to unbalanced radial force. Therefore, the radial clearance may be slightly larger, which is generally 0.13-0.16 mm. For the purpose of preventing the pressure fluid from leaking out of the pump from the position between the pump body and the pump cover, and reducing the tension of the forcing screw, a fluid seal drain pan 16 is provided on the end face on both sides of the pump body, so as to introduce the pressure fluid permeating between the pump body and pump cover into the fluid suction chamber. The small holes on the pump cover and driven shaft are used for introducing the pressure fluid leaked to the bearing end into the fluid suction chamber, to avoid fluid overflowing and lubricate the needle bearing.

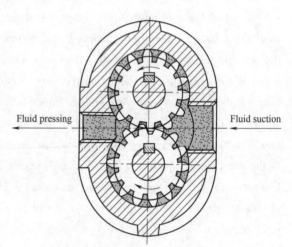

Figure 1-2-2 Working Principle of Externally-engaged Gear Pump

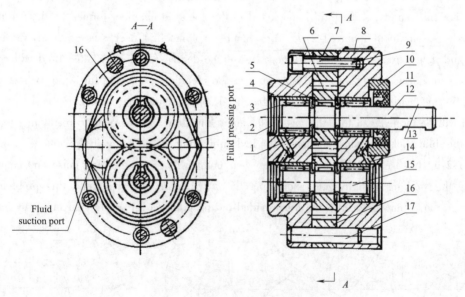

Figure 1-2-3 Structure of Gear Pump

1-Outer race of bearing; 2-Plug; 3-Roller; 4-Rear pump cover; 5-Key; 6-Gear;
7-Pump body; 8-Front pump cover; 9-Screw; 10-Compression ring; 11-Seal ring; 12-Driving shaft;
13-Key; 14-Fluid drain hole; 15-Driven shaft; 16-Drain pan; 17-Positioning pin

Task 2 Knowledge about Gear Pump

🔍 Task description

What are the working principle, advantages and disadvantages of the gear pump? What problems will appear in the process of use? What are the main application scenarios of gear pumps?

🧩 Relevant knowledge

The gear pump shown in Figure 1-2-1 looks like a water pump from its appearance, but actually, it is like a lift pump in terms of working property, which sucks fluid out of the tank. The gear pump is a hydraulic pump that is mostly used in the hydraulic system, featuring simple structure, convenient manufacturing, low price, small volume, light weight, good self-priming performance, insensitivity to fluid pollution and reliable operation. Its main disadvantages are high flow and pressure fluctuation, loud noise and non-adjustable displacement.

Figure 1-2-1 Gear Pump

I Working Principle and Structure of Gear Pump

There is a pair of externally-engaged gears in the gear pump body. End covers are provided on both sides of the gears. Various enclosed working chambers are formed among the pump body, end cover and various gears. If the gear rotates in a direction shown in Figure 1-2-2, the volume of the enclosed working chamber (of the right fluid suction chamber) will increase as the engaged gear teeth gradually disengage, forming a partial vacuum. Under the action of atmospheric pressure, the fluid in the tank will be forced from the tank to fill up the tooth space, and brought to the left fluid pressing chamber due to gear rotation. For the fluid pressing area, the volume of the enclosed working chamber will decrease continuously as the gear teeth gradually engage, so as to force the fluid out.

The principle of gear pump is shown in Figure 1-2-2, and the structure in Figure 1-2-3. The front and rear covers of the pump and the pump body are positioned by two positioning pins and fastened with 6 screws. To ensure that the gear can rotate flexibly with minimum leakage, an appropriate clearance (axial clearance) should be provided between the gear end face and the pump cover. The gap should be 0.025-0.04 mm for small flow-pumps, and 0.04-0.06 mm for large-flow pumps. The clearance (radial clearance) between the addendum and the inner surface of the pump body has little influence on leakage because the sealing belt is long and the shear flow formed by the

$$\eta_m = \frac{T_i}{T_0} = \frac{1}{1+\frac{\Delta T}{T_1}} \qquad (1\text{-}5)$$

(2) Power of hydraulic pump:

①Input power P_i: mechanical power acting on the hydraulic pump spindle. When the input torque is T_0 and the angular velocity is ω

$$P_i = T_0 \omega \qquad (1\text{-}6)$$

②Output power P_o: product of pressure difference Δp between actual fluid suction and pressing ports and output flow q when the hydraulic pump is working

$$P_o = \Delta p \cdot q \qquad (1\text{-}7)$$

Where: Δp is the pressure difference (in N/m^2) between actual fluid suction and pressing ports, q is the actual output flow rate (in m^3/s), and P_o is the output power of the hydraulic pump (in $N \cdot m/s$ or W).

In the actual calculation, if the fluid tank is open to the atmosphere, the pressure difference between the fluid suction and pressing ports will often be substituted by the outlet pressure p of the hydraulic pump.

(3) Total efficiency of hydraulic pump: the ratio of the actual output power of the hydraulic pump to the input power, namely

$$\eta = \frac{P}{P_i} = \frac{\Delta p \cdot q}{T_0 \omega} = \frac{\Delta p \cdot q_i \cdot \mu_v}{\dfrac{T_i \omega}{\eta_m}} = \eta_v \eta_m \qquad (1\text{-}8)$$

Where: $\Delta p q_i / \omega$ is the theoretical input torque T_i.

According to the formula (1-8), the total efficiency of the hydraulic pump is the product of the volumetric efficiency and mechanical efficiency, so the input power of the hydraulic pump may also be calculated by

$$P_i = \frac{\Delta p \cdot q}{\eta} \qquad (1\text{-}9)$$

Task implementation

1 Look up the main parameters of the hydraulic pump online, and list them in a table.

2 Grouping and discussion: groups of 3-5. Each student should explain the working principle of the hydraulic positive displacement pump briefly.

3 Look up the materials regarding the development status of hydraulic pumps in China, objectively evaluate them, and put forward opinions.

geometric dimensions of the hydraulic pump enclosed volume per revolution. The hydraulic pumps with variable displacement are called variable displacement pumps, and those with fixed displacement are called fixed-displacement pumps.

(2) Theoretical flow rate q_i: the average volume of fluid discharged per unit time on the premise that the leakage flow rate of the hydraulic pump is not considered. If the hydraulic pump displacement is V and its spindle speed is n, the theoretical flow rate q_i of the hydraulic pump is

$$q_i = Vn \qquad (1\text{-}1)$$

(3) Actual flow rate q: the volume of fluid discharged by the hydraulic pump per unit time under a specific working condition. It is equal to the theoretical flow rate q_i minus the leakage flow rate Δq, namely

$$q = q_i - \Delta q \qquad (1\text{-}2)$$

(4) Rated flow rate q_n: the flow rate of the hydraulic pump that must be guaranteed under normal working conditions according to the test standards (e.g. under rated pressure and rated speed).

3 Power and efficiency

(1) Power loss of hydraulic pump: It is divided into volumetric loss and mechanical loss.

①Volumetric loss: loss of hydraulic pump flow. The actual output flow of the hydraulic pump is always less than the theoretical flow. The causes mainly include the leakage of the high-pressure chamber in the hydraulic pump, fluid compression, and insufficient fluid filled in the enclosed working chamber due to the high fluid suction resistance, fluid viscosity, or hydraulic pump speed during fluid suction. The volumetric loss of the hydraulic pump is expressed by volumetric efficiency η_i (the ratio of the actual output flow rate q to the theoretical flow q_i)

$$\eta_i = \frac{q}{q_i} = \frac{q_i - \Delta q}{q_i} = 1 - \frac{\Delta q}{q_i} \qquad (1\text{-}3)$$

As a result, the actual output flow rate q of the hydraulic pump is

$$q = q_i \eta_v = Vn \eta_v \qquad (1\text{-}4)$$

Where: V is the hydraulic pump displacement (in m^3/r) and n hydraulic pump speed (in r/s).

The volumetric efficiency of the hydraulic pump decreases as the working pressure of the hydraulic pump increases, and varies with the structure types of the hydraulic pump, but it is always less than 1.

②Mechanical loss: torque loss of the hydraulic pump. The actual input torque T_0 of the hydraulic pump is always greater than the theoretically required torque T_i. The main causes include the friction torque loss resulting from the mechanical friction between the relatively moving parts in the hydraulic pump body and the friction loss due to fluid viscosity. The mechanical loss of the hydraulic pump is expressed by mechanical efficiency (the ratio of theoretical torque T_i to the actual input torque T_0). If the torque loss is ΔT, the mechanical efficiency of the hydraulic pump will be

during fluid suction. The pressure of the suction chamber depends on the fluid suction height and the resistance of the fluid suction pipeline. If the height or the resistance is excessively high, the vacuum degree of the chamber will be too high to affect the self-priming capacity of the hydraulic pump. The chamber is called fluid pressing chamber during fluid pressing. The pressure of the fluid pressing chamber depends on external loads and the pressure loss of the fluid discharge pipeline. Theoretically speaking, the fluid discharge pressure has nothing to do with the hydraulic pump flow rate.

The theoretical flow rate of the hydraulic positive displacement pump is dependent on the relevant geometric dimensions and speed of the hydraulic pump. It is independent of the fluid discharge pressure. However, the fluid discharge pressure will affect the pump internal leakage and fluid compression, thus influencing the actual output flow rate of the pump. Accordingly, the actual output flow rate of the hydraulic pump decreases as the fluid discharge pressure increases.

The hydraulic pumps can be divided into fixed-displacement pumps and variable displacement pumps according to whether the volume of fluid they output per unit time can be adjusted. They classified by structure into the gear, vane and plunger types. The graphic symbols of hydraulic pumps are shown in Figure 1-1-2.

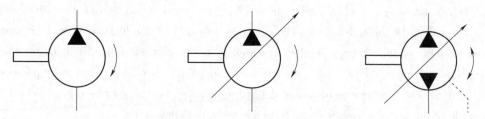

(a) One-way fixed-displacement motor   (b) One-way variable displacement pump   (c) Two-way variable displacement pump

Figure 1-1-2   Graphic Symbols of Hydraulic Pumps

II Main Performance Parameters of Hydraulic Pump

1 Pressure

(1) Working pressure: output pressure of the hydraulic pump during actual operation. The working pressure depends on the external loads and the pressure loss of the fluid discharge pipeline. It has nothing to do with the hydraulic pump flow rate.

(2) Rated pressure: the maximum pressure at which the hydraulic pump, according to the test standard, can run continuously under normal working conditions.

(3) Maximum allowable pressure: the maximum pressure (greater than the rated pressure) at which the hydraulic pump can run for a short time according to the test standard.

2 Displacement and flow rate

(1) Displacement V: the volume of the discharged fluid calculated based on the change in the

always pressed against the eccentric wheel 1. The prime mover drives the eccentric wheel 1 to rotate and the plunger 2 to reciprocate so that the enclosed volume a changes in a periodical and alternate manner. If the volume a increases, a partial vacuum will be formed. Under the action of atmospheric pressure, the check valve 6 will be opened by the fluid via the fluid suction pipe, and then the fluid will enter the chamber a to realize fluid suction. Conversely, if the volume a decreases, the check valve 5 will be opened by the fluid sucked in the chamber a, and then the fluid will flow into the hydraulic system to realize fluid pressing. In this way, the hydraulic pump is available for continuous fluid suction and pressing.

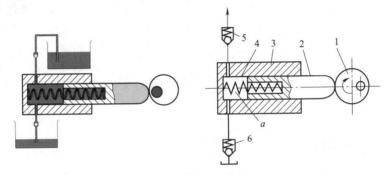

Figure 1-1-1　Working Principle Diagram of Hydraulic Single Plunger Pump

1-Eccentric wheel; 2-Plunger; 3-Cylinder body; 4-Spring; 5,6-Check valve

2　Characteristics of Hydraulic Pump

The hydraulic single plunger pump has all basic characteristics of the hydraulic positive displacement pump:

(1) There are several enclosed spaces that can change periodically. The output flow rate of the hydraulic pump is in direct proportion to the volume change of this space and the number of changes per unit time. It is independent of other factors. This is an important characteristic of the hydraulic positive displacement pump.

(2) The absolute pressure of the fluid in the tank must be identically equal to or greater than atmospheric pressure. This is an external condition for a hydraulic positive displacement pump to suck fluid. Therefore, the tank must be open to the atmosphere, or a sealed pressurized tank must be adopted, so as to ensure that the hydraulic pump can suck fluid normally.

(3) A corresponding fluid distribution mechanism should be provided to separate the fluid suction chamber from the fluid discharge chamber, thus ensuring that the hydraulic pump can suck and discharge the fluid in a regular and continuous way. The structural principle of the hydraulic pump is different, so is the fluid distribution mechanism. Check valves 5 and 6 shown in Figure 1-1-1 are fluid distribution mechanisms.

The fluid chamber in the hydraulic positive displacement pump is called fluid suction chamber

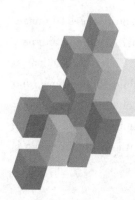

Part 1
Knowledge about Hydraulic Power Elements

Task 1 Overview of Hydraulic Power Elements

Task description

The power source of the hydraulic system is the hydraulic pump. This task focuses on their characteristics, and the working principle of the hydraulic pump. After learning, students are required to understand some detailed parameters of the hydraulic pump.

Relevant knowledge

I Working Principle and Characteristics of Hydraulic Pump

The hydraulic power element, serving as the power source of the whole hydraulic system, is a core element of the system. The most commonly used hydraulic power element is the hydraulic pump, which is an energy conversion device for converting the mechanical energy output by a prime mover (motors or internal combustion engines are the most common ones) into the pressure energy of working fluid (hydraulic energy).

Driven by the prime mover, the hydraulic pump converts the input mechanical energy into the pressure energy of fluid, and then inputs it into the system in the form of pressure and flow. It is the heart of hydraulic transmission and the power source of the hydraulic system.

1 Working principle of hydraulic pump

Generally, hydraulic pumps work by virtue of the change in the enclosed volume. Therefore, they are referred to as hydraulic positive displacement pumps. Figure 1-1-1 shows the working principle of a single plunger hydraulic pump. As shown in the figure, the plunger 2 is installed in the cylinder 3 to form the enclosed volume a, and under the action of the spring 4, the plunger is

Hydraulic and Pneumatic Technology

Part 5 Pneumatic Transmission Fundamentals ········ 097
Task 1 Working Principle and Composition of Pneumatic Transmission System ······ 097
Task 2 Characteristics of Pneumatic Transmission System ················ 099
Task 3 Applications of Pneumatic Transmission System ·················· 101
Thinking and Exercise ··· 103

Part 6 Knowledge about Pneumatic Components ········ 104
Task 1 Air Supply Device ······································ 104
Task 2 Knowledge about Pneumatic Actuating Components ················ 108
Task 3 Knowledge about Pneumatic Control Components ················· 113
Thinking and Exercise ··· 121

Part 7 Pneumatic Circuits ··························· 122
Task Installation and Commissioning of Basic Pneumatic Circuits ············· 122
Thinking and Exercise ··· 128

Part 1 Knowledge about Hydraulic Power Elements ········ 001

Task 1 Overview of Hydraulic Power Elements ········ 001
Task 2 Knowledge about Gear Pump ········ 006
Task 3 Knowledge about Vane Pump ········ 010
Task 4 Knowledge about Plunger Pump ········ 017
Task 5 Selection of Hydraulic Pump ········ 021
Thinking and Exercise ········ 023

Part 2 Hydraulic Actuators ········ 024

Task 1 Knowledge about Hydraulic Motor ········ 024
Task 2 Knowledge about Hydraulic Cylinder ········ 031
Thinking and Exercise ········ 041

Part 3 Hydraulic Control Valve ········ 042

Task 1 Knowledge about Directional Control Valve ········ 042
Task 2 Knowledge about Pressure Control Valve ········ 054
Task 3 Knowledge about Flow Control Valve ········ 066
Task 4 Knowledge about Sandwich Valve ········ 068
Thinking and Exercise ········ 071

Part 4 Installation and Commissioning of Basic Hydraulic Circuits ········ 073

Task 1 Installation and Commissioning of Directional Control Circuits ········ 073
Task 2 Installation and Commissioning of Pressure Control Circuits ········ 075
Task 3 Installation and Commissioning of Speed Control Circuits ········ 080
Task 4 Installation and Commissioning of Sandwich Valve Control Circuits ········ 086
Task 5 Installation and Commissioning of Proportional Valve Control Circuits ········ 092
Thinking and Exercise ········ 096

the cultivation of technical application ability and the training of practical technology application.

(2) The content of the teaching material has focused on the analysis of working principles, structures, and applications to the hydraulic and pneumatic equipment.

(3) The hydraulic and pneumatic function symbols in the teaching material have strictly complied with international standards.

(4) The content of the teaching material has demonstrated the complex structural schematic diagram in 3D animation, which will be easier for students to understand.

(5) The content of the teaching material has clear objectives for each task in order to improve students' learning interest and efficiency.

There are 7 projects designed in the teaching material, which are: Knowledge about Hydraulic Power Elements; Hydraulic Actuators; Hydraulic Control Valve; Installation and Commissioning of Basic Hydraulic Circuits; Pneumatic Transmission Fundamentals; Knowledge about Pneumatic Components; and Pneumatic Circuits. Teachers can manage teaching content according to the actual situation of students. Generally speaking, the arrangement of class hours is to meet the needs of students' learning knowledge and cultivate their practical ability.

The chief editors of the teaching material are Gong Xiaokai and Liu Enyu of Tianjin Vocational College of Mechanics and Electricity. The specific division of editing is as follows: tasks one to four are edited by Gong Xiaokai; tasks five to seven are edited by Liu Enyu. Zhejiang Tianhuang Science & Technology Industrial Co., Ltd. offered their great support and help in the process of compiling the teaching material. Here, we would like to express our heartfelt gratitude to them.

In case there has any shortcomings or mistakes, suggestions for improvement will be gratefully received.

Editors
Feb. 2023

When reviewing the historic achievements and changes that have taken place in our country in the new era past ten years, the report to 20th CPC National Congress: "We have pursued a more proactive strategy of opening up. We have worked to build a globally-oriented network of high-standard free trade areas and accelerated the development of pilot free trade zones and the Hainan Free Trade Port. As a collaborative endeavor, the Belt and Road Initiative has been welcomed by the international community both as a public good and a cooperation platform." In order to promote the development of "the Belt and Road" high quality, and expand the cooperation in the area of vocational education with the countries along the Belt and Road, and implement the requirements of the plan of Tianjin that sharing the launched and implemented high quality vocational education results with the world, the vocational education is playing an important role as an education form which is closely related to the manufacturing industry.

Tianjin Vocational College of Mechanics and Electricity established the "Luban Workshop" in Madagascar. In order to cooperate with theory and training teaching of Madagascar's "Luban Workshop", carry out exchange and cooperation, realize educational resource sharing and improve international influence of Chinese vocational education, innovate the international cooperation mode of vocational colleges, and export China's high quality vocational resources, the research group compiled the textbook *Hydraulic and Pneumatic Technology*.

Carrying the spirit of hydraulic equipment of Luban workshop, the teaching material is based on project teaching, serving the vocational ability training of equipment manufacturing specialty in higher vocational colleges. The teaching material is rich in content, detailed in information, and novel in theme. Aiming at cultivating the students' practical ability, the combination of practice and theory, and the analysis and special implementing scheme on solving practical problems are focused, with scientific and advanced nature, the following characteristics is highlighted:

(1) The content of the teaching material cited teaching has cased close to engineering practice considering the post characteristics of higher vocational education talents, focused on

Brief Introduction of Contents

The main contents of this teaching material include hydraulic components, hydraulic basic circuits, pneumatic components, pneumatic basic circuits and their applications. A total of 7 projects and 23 learning tasks are designed with clear objectives in order to improve learning interest and efficiency for the students. Aiming at cultivating the students' practical ability comprehensively, the combination of practice and theory is highlighted, and the analysis and special implementing scheme on solving practical problems are focused, with a certain scientific and advanced nature.

This is a practical teaching material suitable for teachers and students of mechanical automation specialty in higher vocational colleges, and can also be used as a reference book for the training and teaching of hydraulic and pneumatic technology.

The Construction Achievements of National Modern Vocational Education Reform and Innovation Demonstration Area
The R&D Achievements of National Vocational Education Quality Development Research Center
Planned Textbook of the Teaching Mode of Engineering Practice Innovation Project (EPIP)

Hydraulic and Pneumatic Technology

Chief Editors © Gong Xiaokai, Liu Enyu

中国铁道出版社有限公司
CHINA RAILWAY PUBLISHING HOUSE CO., LTD.